# Practical Natural Language Processing with Python

With Case Studies from Industries Using Text Data at Scale

Mathangi Sri

**Apress®**

## Practical Natural Language Processing with Python

Mathangi Sri
Bangalore, Karnataka, India

ISBN-13 (pbk): 978-1-4842-6245-0                    ISBN-13 (electronic): 978-1-4842-6246-7
https://doi.org/10.1007/978-1-4842-6246-7

Managing Director, Apress Media LLC: Welmoed Spahr
Acquisitions Editor: Celestin Suresh John
Development Editor: Matthew Moodie
Coordinating Editor: Aditee Mirashi

Cover designed by eStudioCalamar

Cover image designed by Freepik (www.freepik.com)

Distributed to the book trade worldwide by Springer Science+Business Media New York, 1 New York Plaza, Suite 4600, New York, NY 10004-1562, USA. Phone 1-800-SPRINGER, fax (201) 348-4505, e-mail orders-ny@springer-sbm.com, or visit www.springeronline.com. Apress Media, LLC is a California LLC and the sole member (owner) is Springer Science + Business Media Finance Inc (SSBM Finance Inc). SSBM Finance Inc is a **Delaware** corporation.

For information on translations, please e-mail booktranslations@springernature.com; for reprint, paperback, or audio rights, please e-mail bookpermissions@springernature.com.

Apress titles may be purchased in bulk for academic, corporate, or promotional use. eBook versions and licenses are also available for most titles. For more information, reference our Print and eBook Bulk Sales web page at www.apress.com/bulk-sales.

Any source code or other supplementary material referenced by the author in this book is available to readers on GitHub via the book's product page, located at www.apress.com/978-1-4842-6245-0. For more detailed information, please visit www.apress.com/source-code.

Printed on acid-free paper

*To my loving daughter Sahana*

*who inspires me every day with
her unending spirit to learn*

# Table of Contents

# About the Author

**Mathangi Sri** is a renowned data science leader in India. She has 11 patent grants and 20+ patents published in the area of intuitive customer experience, indoor positioning, and user profiles. She has 16+ years of a proven track record in building world-class data science solutions and products. She is adept in machine learning, text mining, and NLP technologies and tools. She has built data science teams across large organizations like Citibank, HSBC, and GE as well as tech startups like 247.ai, PhonePe, and Gojek. She advises startups, enterprises, and venture capitalists on data science strategy and roadmaps. She is an active contributor on machine learning to many premier institutes in India. She was recognized as one of "The Phenomenal SHE" by the Indian National Bar Association in 2019.

# About the Technical Reviewer

 **Manohar Swamynathan** is a data science practitioner and an avid programmer, with over 14+ years of experience in various data science-related areas, including data warehousing, business intelligence, analytical tool development, ad-hoc analysis, predictive modeling, data science product development, consulting, formulating strategy, and executing analytics programming. His career has covered the life cycle of data across different domains such as US mortgage banking, retail/e-commerce, insurance, and industrial IoT. He's also involved in the technical review of books about data science using Python and R. He has a bachelor's degree with a specialization in physics, mathematics, and computers, and a Master's degree in project management. He's currently living in Bengaluru, the silicon valley of India.

# Acknowledgments

I want to thank my husband, Satish. I brainstormed with him several times during the course of this book on the technical and business use cases. I also want to thank the team at Apress for providing adequate reviews and guidance on the content. I learned all the concepts on the job, so I thank all the people with whom I have had the privilege to work with and learn from.

# Introduction

I am fortunate to have had the exposure and opportunity to solve complex NLP problems that benefited various businesses across geographies. This book comes from my learnings and hence it is a practitioner view of solving text problems. Solving NLP problems involves a certain combination of creativity and technical knowledge. Sometimes the best deep learning methods do not solve a problem as well as simple solutions do. I am always reminded of Occam's razor, which states that when there are alternatives available, the simplest one possibly solves the problem best.

In my opinion, the answer to any problem lies in the data. This is why the first chapter talks about text data. I cover different types of text data and the information that can be extracted from this data. Chapters 2, 3, 4, and 5 focus on different industries and the common text mining problems faced in these domains. As I cover the use cases in these chapters, I also cover the corresponding text mining concepts along with the relevant code and Python libraries. Chapter 2 takes a quick view of the customer service industry with a deep dive into the conversation corpus. It highlights the various types of information that can be extracted from the conversation corpus along with detailed code. I also cover bag of words, vectorization, the rules-based approach, and supervised learning methods in great detail. Chapter 3 focuses on a very popular text mining use case: review mining. I delve into sentiment analysis and the various facets of opinion mining. I cover sentiment analysis using unsupervised and supervised techniques. I introduce neural networks for sentiment analysis in this chapter. Chapter 4 provides an overview of techniques used in the banking and financial services industries. Traditionally, banking uses structured data to drive decisions. Of late, text use cases are being explored in this arena and I talk about one in detail. I explore named entity recognizers in detail in this chapter using unsupervised techniques and embedding based neural networks. Chapter 5 is dedicated to virtual assistants. I explore techniques to build bots using state-of-the-art neural network architectures. This chapter also introduces natural language generation concepts.

# CHAPTER 1

# Types of Data

Natural language processing (NLP) is a field that helps humans communicate with computers naturally. It is a shift from the era when humans had to "learn" to use computers to computers being trained to understand humans. It is a branch of artificial intelligence (AI) that deals with language. The field dates back to the 1950s when a lot of research was undertaken in the machine translation area. Alan Turing predicted that by the early 2000s computers would be able to flawlessly understand and respond in natural language that you won't be able to distinguish between humans and computers. We are far from that benchmark in the field of NLP. However, some argue that this may not even be the right lens to measure achievements in the field. Be that as it may, NLP is central to the success of many businesses. It is very difficult to imagine life without Google search, Alexa, YouTube recommendations, and so on. NLP has become ubiquitous today.

In order to understand this branch of AI better, let's start with the fundamentals. Fundamental to any data science field is data. Hence understanding text data and various forms of it is at the heart of performing natural language processing. Let's start with some of the most familiar daily sources of text data, from the angle of commercial usage:

- Search
- Reviews
- Social media posts/blogs
- Chat data (business-to-consumer and consumer-to-consumer)
- SMS data
- Content data (news/videos/books)
- IVR utterance data

© Mathangi Sri 2021
M. Sri, *Practical Natural Language Processing with Python*, https://doi.org/10.1007/978-1-4842-6246-7_1

# Search

Search is one of the most widely used data sources from a customer angle. All search engine searches, whether a universal search engine or a search inside a website or an app, use at the core indexing, retrieval, and relevance-ranking algorithms. Search, also referred to as a query, is typically made up of short sentences of two or three words. Search engine results are approximate and they don't necessarily need to be bang on with their results. For a query, multiple options are always presented as results. This user interface transfers the onus of finding the answer back to the user. Recount the number of times you have modified your query because you were not satisfied with the result. It's unlikely that you blamed the performance of the engine. You focused your attention on modifying your query.

# Reviews

Reviews are possibly the most widely analyzed data. Since this data is available openly or is easy to extract with web crawling, many organizations use this data. Reviews are very free flowing in nature and are very unstructured. Review mining is core to e-commerce companies like Amazon, Flipkart, eBay, and so on. Review sites like IMDB and Tripadvisor also have reviews data at their core. There are other organizations/ vendors that provide insights on reviews collected by these companies. Figure 1-1 shows sample review data from `www.amazon.in/dp/B0792KTHKK/ref=gw-hero-PC-dot-news-sketch?pf_rd_p=865a7afb-79a5-499b-82de-731a580ea265&pf_rd_r=TGGMS83TD4VZW7KQQBF3`.

Amit
TOP 500 REVIEWER
*3.0 out of 5 stars*
Good <u>speaker</u>, Bad <u>software</u>
20 October 2018
Colour: GreyConfiguration: Echo Dot (single pack)Verified Purchase
This is very good hardware and speaker. However very much handicapped by amazon on software front. It can not understand your commands 70-80 percentage times and very limited to amazon music and saavn. However you can not use any of good music apps like <u>wynk, gaana, youtube etc</u>.
Also natural language processing is supported in google products but not in this echo.
For example when you say-
1) where is agra - it gives you correct answer.
2) continue with next query like how to reach there ?
Then for 2nd query it doesn't know what to do..
This is just an example but most of the times you will find this issue there.

Then you can not connect with other populer music etc apps. Have to rely on what amazon thinks os good for you.
Amazon music has very limited music base.

So looking forward to get more access and integration with alexa however not have much hope since its already such an old platform that they would have given support if there were any such plans.

*Figure 1-1.  Sample Amazon review*

Note that the above review highlights the features that are important to the user: the scope of the product (music), the search efficiency, the speaker, and its sentiment. But we also get to know something about the user, such as the apps they care about. We could also profile the user on how objective or subjective they are.

As a quick, fun exercise, look at the long review from Amazon in Figure 1-2 and list the information you can extract from the review in the following categories: product features, sentiment, about the user, user sentiment, and whether the user is a purchaser.

DANNEL

TOP 500 REVIEWER

*4.0 out of 5 stars*

Brilliant Machine - Updated 26/06/2018 - 24 Months Review!

15 June 2016

Verified Purchase

This is my first front loader. I've used it for a while and I can tell you this is awesome! It is super silent and the vibrations are very minimal if not hard to notice. It has all the washing modes a front loader can have. In case you are curious, the Monsoon mode is used for washing rain soaked clothes that smell. I particularly liked the Drum Clean mode. It makes it a bit easier to maintain the machine. The size is just right. The grey color looks great! The quality of the machine is superb!

Using a trolley will allow you to move this machine whenever you need to clean the floor area of the machine or to make adjustments to the machine when issues come up. Besides it protects the machine from external water leakages. But it is important to use a quality stand for this purpose or else you risk damaging the machine through excessive vibrations! Do note that Bosch doesn't recommend a stand as it will void the warranty. If you have voltage problems, it's a must to install a stabilizer or spike guard. Installing a water filter is important as some particles may enter the machine making it difficult to maintain it. Always use a tap connection that has a water pressure up to 8 liters per minute. The rubber seal on the entrance needs to be wiped clean to prevent mold formations. If necessary, use white vinegar to clean the seal as it is helpful in removing odor and mold formations. Keep the door open for a few hours once the washing is done. This allows the moisture to dry within the drum. Finally, follow your manufacturer's instructions on the machine's usage and maintenance. This will lengthen the life of your machine.

After a lot of research, I've found that in general, the technology used in all front loaders including Bosch isn't perfect. In fact, several cases are pending against front load machine manufacturers in the U.S. for continuing to use this imperfect technology as this has not made consumer's life easier. Now you might wonder what this is all about! It's all about maintenance! There are many issues that are caused by manufacturing defects. These front loaders get mold formation easily plus some people have even reported drum issues like rusting! Improper maintenance on the part of the owners will also lead to these issues. Beware that the Printed Circuit Board (PCB) can be damaged by rough or improper handling of the machine's controls (Buttons & Dial). Power spikes and low voltages in India are the biggest reasons why majority of the owners are affected by burnt PCBs. Replacing the PCB is very expensive (Costs Rs. 10,000 in India!). This is seriously high maintenance! The deal breaker is that it is cheaper to buy a new machine instead of repairing the existing machine (in case of a drum issue)! What I am trying to say is that you must be very careful on the maintenance aspect.

***Figure 1-2.*** *Extract some data from this review.*

# Social Media Posts/Blogs

Social media posts and blogs are widely researched, extracted, and analyzed, like reviews. Tweets and other microblogs are short and hence could seem easily extractable. However, tweets, depending on use cases, can carry a lot of noise. From my experience, on average only 1 out of every 100 tweets contains useful information on a given concept of interest. This is especially true in cases of analyzing sentiments for brands using Twitter data. In this research paper on sentiment analysis, only 20% of tweets in English and 10% of tweets in Turkish were found to be useful after collecting tweets for the topic: www.researchgate.net/profile/Serkan_Ayvaz/publication/320577176_Sentiment_ Analysis_on_Twitter_A_Text_Mining_Approach_to_the_Syrian_Refugee_Crisis/ links/5ec83c79299bf1c09ad59fb4/Sentiment-Analysis-on-Twitter-A-Text-Mining- Approach-to-the-Syrian-Refugee-Crisis.pdf. Hence looking for the right tweet in a corpus of tweets is a key to successfully mining Twitter or Facebook posts. Let's take an example from https://twitter.com/explore:

```
Night Santa Cruz boardwalk and ocean
Took me while to get settings right. .....
Camera: pixel 3
Setting: raw, 1...https://t.co/XJfDq4WCuu
@Google @madebygoogle could you guys hook me up with the upcoming Pixel
4XL for my pixel IG. Just trying to stay ah...https://t.co/LxBHIRkGG1
China's bustling cities and countryside were perfect for a smartphone
camera test. I pitted the #HuaweiP30Pro again...https://t.co/Cm79GQJnBT
#sun #sunrise #morningsky #glow #rooftop #silohuette madebygoogle google
googlepixel #pixel #pixel3 #pixel3photos...https://t.co/vbScNVPjfy
RT @kwekubour: With The Effortlessly Fine, @acynam
📷x Pixel 3
Get A #Google #Pixel3 For $299, #Pixel3XL For $399 With Activation In
These Smoking Hot #Dealshttps://t.co/ydbadB5lAn via @HotHardware
I purchased pixel 3 on January 26 2019 i started facing call drops issue
and it is increasing day by day.i dont kn...https://t.co/1LTw9EdYzp
```

As you can see in this example, which displays sample tweets for Pixel 3, the content spans deals, reviews of the phone, amazing shots taken from the phone, someone awaiting the Pixel 4, and so on. In fact, if you want to understand the review or sentiment associated with Pixel 3, only 1 out of the 8 tweets is relevant.

A microblog's data can contain power-packed information about a topic. In the above example of Pixel 3, you can find the following: the most liked or disliked features, the influence of location on the topic, the perception change over time, the impact of advertisements, the perception of advertisements for Pixel 3, and what kind of users like or dislike the product. Twitter can be mined as a "leading indicator" for various events, such as if a stock price of a particular company can be predicted if there is significant news about the company. The research paper at www.sciencedirect.com/science/ article/pii/S2405918817300247 describes how Twitter data was used to correlate the movements of the FTSE Index (Financial Times Stock Exchange Index, an index of the 100 most capitalized companies on the London Stock Exchange) before, during, and after local elections in the United Kingdom.

# Chat Data

## Personal Chats

Personal chats are the classic everyday corpus of WhatsApp chat or Facebook or any other messenger service. They are definitely one of the richest sources of information to understand user behavior, more in the friends-and-family circle. They are filled with a lot of noise that needs to be weeded out, like you saw with the Twitter data. Only a small portion of the corpus is relevant for extracting useful commercial information. The incidence rate of this commercially useful information is not very high. That is to say, it has a low signal-to-noise ratio. The paper at `www.aaai.org/ocs/index.php/ICWSM/ICWSM18/ paper/view/17865/17043` studied openly submitted data from various WhatsApp chats. Figure 1-3 shows a word cloud of the WhatsApp groups analyzed by the paper.

***Figure 1-3.*** *WhatsApp word cloud*

Also the data privacy guidelines of messenger apps may not permit mining personal chats for commercial purposes. Some of the personal chats have functionality where a business can interact with the user, and I will cover that as part of the next section

# Business Chats and Voice Call Data

Business chats, also referred to as live chats, are conversations that consumers have with a business on their website or app. Typically users reach out to chat agents about the issue they face in using a product or service. They may also discuss the product before making a purchase decision. Business chats are fairly rich in information, more so on the commercial preferences of the user. Look at the following example of a chat:

| | |
|---|---|
| system | Thank you for choosing Best Telco. A representative will be with you shortly. |
| system | You are now chatting with 'Max' |
| customer | hi max |
| agent | Thank you for contacting Best Telco. My name is Max. |
| agent | Hello, I see that I'm chatting with Mrs. Sara and you've provided XXXXXXX as the number associated with your account. Is that correct? |
| customer | yes. that is correct.  I just received email saying my service price was going up to $70/month but on your web site, it's only $30. What's the deal with that? |
| agent | Thank you for the information Mrs. Sara. |
| agent | I'm sorry to hear about the increase, our billing department can verify on the changes of the prices. |
| customer | who do I need to talk to? |
| agent | I will provide you the billing department number. Sorry |
| customer | thank you |
| Agent | The number for Billing is: XXXXXXXX, they are open Monday - Friday: 8:00 a.m. - 6:00 p.m. ET. Saturday/Sunday/Holidays: Closed |
| Customer | ok thanks. good bye |

A lot of information can be cleanly taken from the above corpus. The user name, their problem, the fact that the user is responsive to emails, the user is also sensitive to price, the user's sentiment, the courtesy of the agent, the outcome of the chat, the resolution provided by the agent, and the different departments of Best Telco.

Also, note how the data is laid out: it's free flow text from the customer. But the chat agent plays a critical role in directing the chat. The initial lines talk about the issue, then the agent presents a resolution, and then towards the end of the chat a final answer is received along with the customer expressing their sentiment (in this case, positive).

The same interaction can happen over a voice call where a customer service representative and a user interact to solve the issue faced by the customer. Almost all characteristics are the same between voice calls and chat data, except in a voice call the original data is an audio file, which is transcribed to text first and then mined using text mining. At the end of the customer call, the customer service representative jots down the summary of the call. Referred to as "agent call notes," these notes are also mined to analyze voice calls.

# SMS Data

SMS is the best way to reach 35% of the world. SMS as a channel has one of the highest open rates (number of people who open the SMS message to number of people who received the SMS message): 5X over email open rates (`https://blog.rebrandly.com/12-sms-text-message-marketing-statistics`). On average, a person in the US receives 33 messages a day (`www.textrequest.com/blog/how-many-texts-people-send-per-day/`). Many app companies access customers' SMS messages and mine the data to improve user experiences. For instance, apps like ReadItToMe read any SMS messages received by users while they are driving. Truecaller reads the SMS messages and classifies them into spam and non-spam. Walnut provides a view of users' spending based on the SMS messages they have received. Just by looking at only transactional SMS data, much user information can be extracted: user's income, their spending, type of spending, preference for online shopping, etc. The data source is more structured if we are only analyzing business messages. See Figure 1-4.

*Figure 1-4.* *A screenshot from the Walnut (https://capitalfloat.com/walnut/) app*

Businesses follow a template and are more structured. Take the following SMS as an example. The noise in this dataset is much less. Clear information is presented in a clear style. Although different credit card companies can present different styles of information, it is still easier to extract information as compared to free-flow customer text.

```
Mini Statement for Card ******1884.Total due Rs. 4813.70. Minimum due
Rs.240.69. Payment due on 07-SEP-19. Refer to your statement for more
details.
```

# Content Data

There is a proliferation of digital content in our lives. Online news articles, blogs, videos, social media, and online books are key types of content that we consume every day. On average, a consumer spends 8.8 hours consuming content digitally, per `https://cmo.adobe.com/articles/2019/2/5-consumer-trends-that-are-shaping-digital-content-consumption.html`. The following are the key problems data scientists need to solve to use text mining:

- Content clustering (grouping similar)

- Content classification

- Entity recognition

- Analyzing user reviews on content

- Content recommendation

The other key data, like a user's feedback on the content itself, is more structured: number of likes, shares, clicks, time spent, and so on. By combining the user preference data with the content data, we can understand a lot of information about the preference of the user, including lifestyle, life stage, hobbies, interests, and income level.

# IVR Utterance Data

IVR is interactive voice response. IVR is the system that helps you navigate various menu options in a voice call. IVR is estimated to be a $5.5 BN market by 2023, per `www.marketsandmarkets.com/Market-Reports/interactive-voice-response-market-33851149.html?gclid=CjOKCQjwh8jrBRDQARIsAH7BsXfkISk9rnRcUPrjpBP6qC_a3-xBFCaNzEVz4FANz5y__ZXbN7yA9ZoaAkovEALw_wcB`. Natural language-based IVR is an advanced system of IVR where a bot-like interface helps a customer find a solution to their issue or problem. An example of a transcript is shown in Figure 1-5.

**IVR BOT**: Hi welcome to XYZ bank. How can I help you?. You can say things like what is my balance, my card did not go through.

**User**: Hey there I am at a retail store and my card is not working.

**IVR BOT**: For verification provide me your card number please.

**User**: XXXXXXXX

**IVR BOT**: Thanks for the verification. Due to long inactivity your card has been deactivated. Can I go ahead and activate your card.

**User**: Yes, please.

**IVR BOT**: Your card is active now. You can go ahead with your transactons.

**User**: Thanks. That's cool.

*Figure 1-5.* *Example IVR transcript*

IVR utterances are typically short. The user knows they are talking to a bot, so they tend to keep the sentences short. Since the utterances are really to the point, we do not get to know other attributes or preferences of the user other than the problem they are facing. It is a much more structured text than the other examples you have seen so far.

# Useful Information from Data

Having described in detail the different types of data and their applications, Figure 1-6 shows a chart that plots the useful information value of each data source and the ease of extracting the data. The useful information value is the value that can be extracted about a user from the data that is useful to an organization. The ease of extracting data is about dealing with the nature of the text corpus itself.

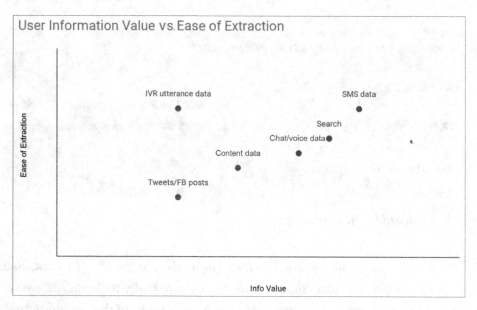

***Figure 1-6.***  *User information value vs. ease of extraction*

# CHAPTER 2

# NLP in Customer Service

Customer service is a multi-billion dollar industry. The cost of bad customer experience is estimated to be a trillion dollars globally (`https://blog.hubspot.com/service/customer-service-stats`). Customer service has its early origins in the form of call centers in the 1960s. The need for customer service grew, and today customer support has become a sizeable portion of any consumer organization. Customers can contact organizations in a multi-modal way, through the Web, apps, voice, IVR, or calls. In this chapter, we will look at the core problems that NLP solves in human-assisted customer service via chats and calls.

Let's look at the data structures in the customer service industry by each of the service channels from which text data is collected.

## Voice Calls

Customers call into contact centers and a customer support agent from the company answers the customers' queries. At the end of the call, the agent notes down the nature of the call, any tickets the agent had raised, and the resolution. The agent may also fill out some forms that pertain to the call. Table 2-1 shows sample data.

© Mathangi Sri 2021
M. Sri, *Practical Natural Language Processing with Python*, https://doi.org/10.1007/978-1-4842-6246-7_2

***Table 2-1.*** *Example of Call Data*

| Call id | Call start | Call end | Agent call notes | Call transferred | Agent name | Customer survey rating | Customer survey comments |
|---|---|---|---|---|---|---|---|
| 424798 | **11/12/2019 1:09:38** | 11/12/2019 1:18:38 | customer asked about high bills. explained. customer fine | No | Sam | | |
| 450237 | **11/26/2019 8:58:58** | 11/26/2019 9:13:58 | waiver fee. escalated | No | Kiran | | |
| 794029 | 4/26/2019 8:23:52 | 4/26/2019 8:34:52 | payment link sent | No | Karthik | 3 | call disconnects.. agent helpful |
| 249311 | 12/8/2019 10:50:14 | 12/8/2019 11:02:14 | xferred to surperv | Yes | Megha | | |

We analyze the agent call notes and customer survey comments. This gets then pivoted with other structured data to derive insights about the calls.

# Chats

Customers can chat with customer support on the e-commerce web page or on mobile apps. You saw an example of a chat transcript in the last chapter. As with calls, the agents are made to fill an "agent disposition form" at the end of an interaction. The agent disposition forms in chats are slightly longer as compared to voice calls. The agents can multitask and hence are expected to fill in details about the chats so we get line-level text data along with time stamps. A small example of this data is shown in Table 2-2.

***Table 2-2.*** *Example of Chat Data*

| Speaker | Line text | Time stamp |
|---|---|---|
| system | Thank you for choosing Best Telco. A representative will be with you shortly. | 5:44:44 |
| system | You are now chatting with Max. | 5:44:57 |
| customer | Himax | 5:45:03 |
| agent | Thank you for contacting Best Telco. My name is Max. | 5:45:14 |
| agent | Hello, I see that I'm chatting with Mrs. Sara and you've provided XXXXXXX as the number associated with your account. Is that correct? | 5:45:26 |
| customer | yes. that is correct. I just received email saying my service price was going up to$70/monthbut on your web site, it's only$30. What's the deal with that? | 5:45:39 |

This data is often analyzed with the following other data:

- Metadata of chats, like average handle time, chats that got transferred to another agent, web page where the chat happened

- Metadata of agents, such as agent's tenure and pervious performance history

- Agent disposition data (as filled in by the agent), such as reason for the user's chat, resolved or unresolved, tickets raised, etc.

- User data such as user preferences, history of user behavior data, history of user contact, etc.

- User survey data such as the user rating at the end of the chat and any user feedback comments

# Tickets Data

Tickets generally get created when the customer calls or chats with an agent. In few cases, one may be created on the Web or the app directly by the customer themselves. Table 2-3 shows a quick look at ticket data. Ticket data may be more structured than the above two channels you just examined. The unstructured text here could be of agent notes and in some cases it will be a pretty concise form of what action was taken, making it less useful.

***Table 2-3.*** *Example of Ticket Data*

| Ticket ID | Category | Subcategory | Agent notes | Department | Status | Date of ticket opened | Date of ticket closed |
|---|---|---|---|---|---|---|---|
| 461970 | Repairs | Replacement | sent to vendor a | Merchandise | Closed | 10/1/2019 11:51:16 | 10/3/2019 11:51:16 |
| 271234 | Bills | Waiver | closed on first call | CS | Re-open | 9/28/2019 1:21:11 | 9/30/2019 1:21:11 |
| 356827 | Payment | Failure | called and closed | Collections | Closed | 12/23/2019 8:06:24 | 12/24/2019 8:06:24 |
| 22498 | Network | Intermittent | bad network | Tech Support | Open | 8/10/2019 11:22:03 | |

# Email data

Email is the most cost-effective channel of customer support and is widely used. Table 2-4 shows a sample of an Enron dataset (`www.cs.cmu.edu/~./enron/`) extracted from Enron CSV files.

*Table 2-4.* *Example of an Email Dataset*

| | Date | From | To | Subject | X-To | X-cc | X-bcc | message_fnl |
|---|---|---|---|---|---|---|---|---|
| 0 | Mon, 14 May 2001 16 | phillip. allen@ enron. com | tim. belden@ enron.com | | Tim Belden <Tim Belden/Enron@ EnronXGate> | | | Here is our forecast |
| 1 | Fri, 4 May 2001 13 | phillip. allen@ enron. com | john. lavorato@ enron.com | Re | John J Lavorato <John J Lavorato/ENRON@ enronXgate@ENRON> | | | Traveling to have a business meeting takes the fun out of the trip. Especially if you have to prepare a presentation. I would suggest holding the business plan meetings here then take a trip without any formal business meetings. I would even try and get some honest opinions on whether a trip is even desired or necessary. As far as the business meetings, I think it would be more productive to try and stimulate discussions across the different groups about what is working and what is not. Too often the presenter speaks and the others are quiet just waiting for their turn. The meetings might be better if held in a round table discussion format. My suggestion for where to go is Austin. Play golf and rent a ski boat and jet ski's. Flying somewhere takes too much time. |

*(continued)*

*Table 2-4.* (*continued*)

| | Date | From | To | Subject | X-To | X-cc | X-bcc | message_fnl |
|---|---|---|---|---|---|---|---|---|
| 2 | Wed, 18 Oct 2000 03 | phillip. allen@ enron. com | leah. arsdall@ enron.com | Re | Leah Van Arsdall | | | test successful. way to go!!! |
| 3 | Mon, 23 Oct 2000 06 | phillip. allen@ enron. com | randall. gay@enron. com | | Randall L Gay | | | Randy, Can you send me a schedule of the salary and level of everyone in the scheduling group. Plus your thoughts on any changes that need to be made. (Patti S for example) Phillip |
| 4 | Thu, 31 Aug 2000 05 | phillip. allen@ enron. com | greg.piper@ enron.com | Re | Greg Piper | | | Let's shoot for Tuesday at 11:45. |

This is a sample of columns from the dataset. The actual dataset has a message id for each message, the source .pst file, the From field, etc.

In the next section, you will now see one of the most common use cases of text mining in customer support: Voice of Customer (VOC). VOC analytics can be a dashboard or a report that provides actionable insights into the pulse of the customer.

# Voice of Customer
## Intent Mining

Intents are what the customer wants to address or solve. This is the reason why the customer is contacting the company or organization. Examples of intent are payment failure, complaints about a high bill, return queries, etc. The simplest and quickest way to understand intents in a corpus is through word clouds. However, intents are typically deeper, and for actionable insights, you want to get an accurate category mapped to each chat or voice interaction. Intents are typically mined in practice through rules or through supervised algorithms. Let's look at each of the processes in detail. We would be looking at a dataset that contains set of questions asked by customers in the airline industry.

Let's explore this data before you proceed any further. See Listing 2-1 and Figure 2-1.

*Listing 2-1.* The Dataset

```
import pandas as pd

df = pd.read_csv("airline_dataset.csv",encoding ='latin1')

df.head()
```

```
df["class"].value_counts()
```

```
login       105
other        79
baggage      76
check in     61
greetings    45
cancel       16
thanks       16
```

| | line | class |
|---|---|---|
| 0 | When can I web check-in? | check in |
| 1 | want to check in | check in |
| 2 | please check me in | check in |
| 3 | check in | check in |
| 4 | my flight is tomm can I check in | check in |

***Figure 2-1.*** *The dataset*

As you can see in Figure 2-1, this dataset has two columns, one containing the sentences and the other the category to which the sentence belongs. The category comes from labelling the data manually. You can see that there are seven categories in all, including "other," which is a catch-all category; any sentence that can't be categorized as any of the six categories is marked as other.

# Top Words to Understand Intents

Given a corpus of sentences, you can understand the top intents by extracting the most frequent words in the corpus. Let's use the NLTK package for getting the top key words. FreqDist in NLTK provides the top distribution of words in a group of words. You first need to convert the panda series of sentences to a concatenated string of words. In Listing 2-2, you use the method get_str_list to do the same. Once you have that in said format, you word tokenize through NLTK and give it to the FreqDist package (Listing 2-3).

*Listing 2-2.*

```
def get_str_list(ser):
    str_all = ' '.join(list(ser))

    return str_all
df["line"] = df["line"].str.lower()
df["line"] = df["line"].str.lstrip().str.rstrip()
str_all = get_str_list(df["line"].str.lower())

words = nltk.tokenize.word_tokenize(str_all)
fdist = FreqDist(words)
fdist.most_common()

[('in', 76),
 ('is', 36),
 ('baggage', 33),
 ('what', 32),
 ('check', 30),
 ('checkin', 30),
 ('login', 30),
 ('allowance', 29),
 ('to', 28),
 ('i', 27),
 ('hi', 26),
 ('the', 25),
 ('my', 22),
 ('luggage', 20),
```

*Listing 2-3.*

```
import nltk
from nltk.probability import FreqDist

def remove_stop_words(words,st):
    new_list = []
    st = set(st)
    for i in words:
        if(i not in st):
            new_list.append(i)
```

```
    return new_list
list_all1 = remove_stop_words(words,st)
##we did not word_tokenize as the returned object is already in a list form
fdist = FreqDist(list_all1)

fdist.most_common()

[('baggage', 33),
 ('check', 30),
 ('checkin', 30),
 ('login', 30),
 ('allowance', 29),
 ('hi', 26),
 ('luggage', 20),
 ('pwd', 18),
 ('free', 17),
 ('username', 15),
```

As you can see in Listing 2-2, a lot of not so useful words like "in," "is," and "what" are coming out as the top frequency words. These words are commonly used in all sentences, so you can't use them to make out any meaningful intents. Such words are called "stop words." A list of stop words is generally available on the Internet. You can also edit these stop words depending on your purposes. In order to avoid this problem we write the next code snippet in Listing 2-3 Next, you will import a package named stop_words. The get_stop_words method plus the language specified lists all the stop words. Your task is to remove the stop words from the given corpus and clean up the text. Once done, you'll again apply the FreqDist method and check the output.

You can see in Listing 2-3 that there are queries related to baggage, check in, login, etc. This matches with three of the top six categories that were labelled.

# Word Cloud

A word cloud provides quick and dirty insights into the data you have. Let's use the data you have and make a word cloud with the set of sentences after removing the stop words. You will use WordCloud (https://pypi.org/project/wordcloud/) and the matplotlib package for the same, as shown in Listing 2-5 and Figure 2-2. But first, install the package using pip install, as shown in Listing 2-4.

***Listing 2-4.*** Installing the Packages

```
!pip install wordcloud
!pip install matplotlib
```

***Listing 2-5.*** Making a Word Cloud

```
from wordcloud import WordCloud
import matplotlib.pyplot as plt
def generate_wordcloud(text):
    wordcloud = WordCloud(
            background_color ='white',relative_scaling = 1,
                        ).generate(text)
    plt.imshow(wordcloud)
    plt.axis("off")
    plt.show()

    return wordcloud
##joining the list to a string as an input to wordcloud
str_all_rejoin = ' '.join(list_all1)

str_all_rejoin[0:42]

'can web check-in ? want check please check'

wc = generate_wordcloud(str_all_rejoin)
```

***Figure 2-2.*** *A word cloud*

In Figure 2-2, you can spot topics such as login, baggage, checkin, cancel, and greetings like "hi." You get five out of the six topics that you set out to discover. That's a great start. However, you can see that certain words are repeated in here, like "hi," "checkin," and "pwd." This is occurring because of collocating words in the word corpus. Collocating words have a high chance of occurring together: "make coffee," "do homework," etc. You can remove collocating words by turning off collocations and hence keeping only single words in the word cloud. See Listing 2-6 and Figure 2-3. You start from the word cloud object that was returned from the generate_wordcloud function.

***Listing 2-6.*** Word Cloud Minus Collocated Words

```
###the wordcloud object that got returned and the relative scores
wc.words_
{'login': 1.0,
 'hi hi': 1.0,
 'baggage': 0.782608695652174,
 'checkin checkin': 0.782608695652174,
 'free': 0.7391304347826086,
 'pwd pwd': 0.6956521739130435,
 'username': 0.6521739130434783,
 'baggage allowance': 0.6521739130434783,
 'password': 0.6086956521739131,
 'flight': 0.5652173913043478,
 'check checkin': 0.4782608695652174,
 'checkin check': 0.4782608695652174,
 'want': 0.43478260869565216,
 'check': 0.391304347826087,
def generate_wordcloud(text):
    wordcloud = WordCloud(
            background_color ='white',relative_scaling = 1,
                        collocations = False,
                        ).generate(text)
    plt.imshow(wordcloud)
    plt.axis("off")
    plt.show()

    return wordcloud
```

```
wc = generate_wordcloud(str_all_rejoin)

wc.words_
{'baggage': 1.0,
 'check': 0.9393939393939394,
 'checkin': 0.9090909090909091,
 'login': 0.9090909090909091,
 'allowance': 0.8787878787878788,
 'hi': 0.7878787878787878,
 'luggage': 0.6060606060606061,
 'pwd': 0.5454545454545454,
```

***Figure 2-3.*** *Refined word cloud*

As you can see in Figure 2-3, you get a word cloud with the collocations removed and you can also see that importance of greetings words has also gone down. The above two methods can provide estimates of top topics. However, they fail to ascertain the topic of a given sentence. They also do not provide the frequency distribution of topics like a manual labelling process does.

## Rules to Classify Topics

Another way to mine the topics is to write rules and classify the sentences. This is a commonly used method and is followed when machine learning can't be trained. Rules are of the form of *If a sentence contains word1 and word2, then it could belong to a topic.* There can be different types of rules, such as *And, Not, And with three words,* and so on. The set of rules are joined by an "or" condition. Tools like QDA Miner, Clarabridge, and SPSS can be used to write and execute powerful rules. Rules can also contain grammatical constructs of a sentence like nouns, adverbs, and so on. In Listing 2-7, Listing 2-8, and Figure 2-4, you write your own text mining rule engine using Python. The rules file contains two sets of rules. One is for a single word being present. If it is present, the sentence gets assigned to that category. For example, if the word "baggage" is present, it gets assigned to the category. The other type of rule is an "and" of two words separated by a window. Based on general practice, it is found that a wide variety of topics can be coved by a single word or a two-word match. For example, the rule *"check" and "in" separated (window) by less than two words* can capture a sentence like "check me in." The column "window" in the rules file with a value of -1 means the window is not applicable. Also note that a "hit" with any rule is sufficient for a sentence to get mapped. From that perspective, all rules are "or"ed with each other.

***Listing 2-7.***  Text Mining Rules

```
import pandas as pd
import re
from sklearn.metrics import accuracy_score

df = pd.read_csv("airline_dataset.csv",encoding ='latin1')
rules = pd.read_csv("rules_airline.csv")

def rule_eval(sent,word1,word2,class1,type1,window):
    class_fnl=""
    if(type1=="single"):
        if(sent.find(word1)>=0):
            class_fnl=class1
        else:
            class_fnl==""
    elif(type1=="double"):
        if((sent.find(word1)>=0) &(sent.find(word2)>=0)):
            if(window==-1):
                class_fnl==class1
            else:

                find_text = word1+ ".*" + word2
                list1 =  re.findall(find_text,sent)
                for i in list1:
```

```
            window_size = i.count(' ')-1
            #print (word1,word2,window_size,sent,window)
            if(window_size<=window):
                class_fnl = class1
                break;

    else:
        class_fnl=""

    return class_fnl
```

***Listing 2-8.***

```
rules.head()
```

|   | word | word2 | class | type | window |
|---|------|-------|-------|------|--------|
| 0 | check in | none | check in | single | -1 |
| 1 | checkin | none | check in | single | -1 |
| 2 | checking in | none | check in | single | -1 |
| 3 | bag | none | baggage | single | -1 |
| 4 | luggage | none | baggage | single | -1 |

***Figure 2-4.***  *Text mining rules output*

The function rule_eval takes the sentence and the parameters of the rules and evaluates if there is a hit. If there is a hit, it assigns a corresponding class to the sentence. For rule type "double," a regular expression of the form word1.* word2 is used. This is to calculate the number of words between two matched words. If that number is below the window specified, then it is considered as a hit. See Listing 2-9 and Figure 2-5.

*Listing 2-9.*

```
df["line"]=df["line"].str.lower().str.lstrip().str.rstrip()

topics_list = []
for i,row in df.iterrows():
    sent = row["line"]
    for j,row1 in rules.iterrows():
        word1 = row1["word"]
        word2 = row1["word2"]
        class1 = row1["class"]
        type1 = row1["type"]
        window = row1["window"]
        class1 = rule_eval(sent,word1,word2,class1,type1,window)

        if(class1!=""):
            break;

    topics_list.append(class1)

df["topics"] = topics_list
df.loc[df.topics=="","topics"]="other"

df.head()
```

| | line | class | topics |
|---|---|---|---|
| 0 | when can i web check-in? | check in | check in |
| 1 | want to check in | check in | check in |
| 2 | please check me in | check in | check in |
| 3 | check in | check in | check in |
| 4 | my flight is tomm can i check in | check in | check in |

*Figure 2-5. Assigning topics*

Once you have the assigned topics from the rules added to the dataset, you can compare the manual labelled class and what was extracted from the rules. You will now measure the accuracy of this exercise by comparing manual labels (class) to the topics assigned (topics) by the rules. You do this by using the accuracy method in scikit-learn. See Listing 2-10.

*Listing 2-10.* The Accuracy Method in Scikit Learn

```
accuracy_score(df["class"], df["topics"], normalize=True, sample_
weight=None)
0.7814070351758794
```

You got a 78% accuracy using some quick rules. However, this number needs to be benchmarked using other metrics such as F score, confusion matrix, and so on, which will be covered in detail in the next section. In order to improve the accuracy further, you need to print out the errors and modify the rules.

# Supervised Learning Using Machine Learning

One of the issues with the rules method is that it's very cumbersome and you may also overfit to the current data so much that you may not know the accuracy of a new dataset. Rules get even more complicated as the number of classes increases. You can use machine learning models to learn the patterns and score the new dataset with a new model. Machine learning learns the pattern in the data from historic data. In your case, you have manually labelled data. Remember the process that you followed earlier in order to improve the model? You printed out the errors as compared to the manual labels and then you analyzed the source of the error to correct the rules. Supervised learning follows a similar approach. It initially predicts an output and then keeps correcting the output with the errors it sees in comparison with manually labelled data until the errors are minimized. You will see the supervised process of text mining in detail below.

## Getting Manually Labelled Data

In the airline dataset, the manually labelled data happens to be the dependent variable class. Manual labelling, also referred as annotations, is a time-consuming activity. However, the cost-benefit of this process is found to be worth the effort. There are online annotation companies (such as Mechanical Turks) that provide labelling services. Sometimes an organization will build a team of annotators with quality control processes for supporting large-scale annotations. There are various tools available for annotations. Examples include GATE (`https://gate.ac.uk/`), LightTag (`www.lighttag.io/`), and TagTog (`www.tagtog.net/`). GATE is an open source text mining tool that has various annotation features. Figure 2-6 shows an example of a screen from GATE. Annotators highlight a certain portion in the text and then annotate what the highlighted topic/word is about.

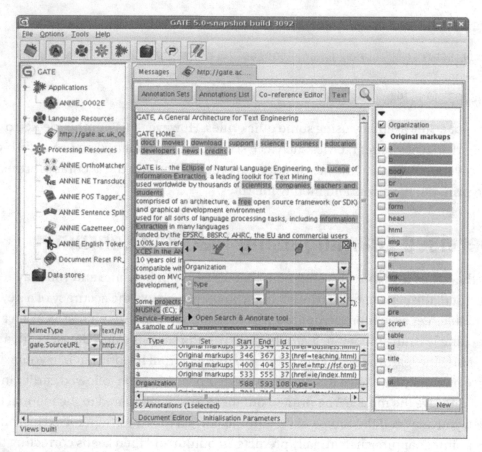

***Figure 2-6.*** *A GATE screen*

However, Microsoft Excel or any Excel-based tool like Google spreadsheets can also help with an annotation problem. Most of the annotations in this book are done through Excel-based platforms including the current case study that you are exploring. Once the sentences or words are labelled, a thorough quality check has to be carried out. Manual labelling accuracies are supposed to be in the high 90% range. This can, of course, change depending on the corpus. As part of quality checks, a supervisor checks the labels done by an annotator. Alternately, labelling accuracies can be increased by getting the corpus labelled by multiple annotators. To get the right quality in the labelling process, the following tasks need to be ensured by the data scientist:

> **Define ontology clearly**: Ontology is the classification tree. In your case, it is the list of topics. They have to be mutually exclusive and collectively exhaustive. This is an important step to achieving high labelling accuracies and hence a high accuracy from the machine learning models.

**Provide examples**: It is also the job of the data scientist to provide examples of each category.

**Conflict between categories**: Sometimes two sentences may seem to belong to different categories. In that case, a clear definition of conflict resolution should be defined. Consider this query: "I am having problems while logging in. I wanted to pay the bill." If there are two categories, Login and Payment, what should this query be labelled?

Once you get the labelled data, the next step is to build a predictive model that can learn patterns in the underlying data. Given a new sentence, it should be able to predict the topic for the same. The set of underlying patterns is referred to as a predictive model. Any predictive model in supervised learning has two components: dependent variables and independent variables. In your case, the dependent variables are the labels and the independent variables are the sentences. However, a machine learning model cannot accept words as inputs. So the words/sentences have to be converted to numbers. The first step in converting a word to numbers is called tokenization.

# Word Tokenization

So you have a bunch of sentences as dependent variables. The first step is to convert them into tokens or units of analysis. Tokens can be words, sentences, or characters. Let's get the tokens for the corpus. See Listing 2-11 to understand ways to tokenize sentences.

*Listing 2-11.* Tokens

```
from nltk.tokenize import word_tokenize

str1 = "What a great show!"
str1.split()
['What', 'a', 'great', 'show!']

from nltk.tokenize import word_tokenize
word_tokenize(str1)
['What', 'a', 'great', 'show', '!']
```

As you can see, the simplest way to break sentences into words is to use the `split` function. However, `split` does not accommodate special characters whereas `word_tokenize` from NLTK does take care of special characters. Sentences can be split by using the `sent_tokenize` function shown in Listing 2-12.

***Listing 2-12.*** Splitting Sentences

```
sentences = "I am feeling great! It was a great show"
from nltk.tokenize import sent_tokenize
sent_tokenize(sentences)
['I am feeling great!', 'It was a great show']
```

Sometimes you may also want to analyze words in a certain order. For example, "check in" means something together and the value of that is lost if it analyzed as "check" and "in." In order to retain as much meaning as possible while converting sentences to word tokens, you also want to retain the order. So instead of having single word tokens, you can have two- or three-term tokens. They are referred as bigrams and trigrams. See Listing 2-13.

***Listing 2-13.*** Bigrams and Trigrams

```
import nltk
text = "I want to check in asap"
bigram_list = list(nltk.bigrams(text.split()))
[' '.join(i) for i in bigram_list]

['I want', 'want to', 'to check', 'check in', 'in asap']
```

Now that you know about tokenization, let's apply it to your corpus and split it into words.

# Term-Document Matrix

Let's now apply word tokenization to the list of sentences you have and get the output. The output will then feed into the machine learning model as the dependent variables. Your objective is to convert the text you have here to numbers. The simplest approach is to use the "bag-of-words" approach. Here you get the list of words or tokens in a corpus and treat them like features in a data mining problem. Now you get a matrix of size N*M where N is the number of documents (length of the dataframe) and M is the unique number of tokens in the corpus. As a start, the cells can be a presence or an absence of words. Use the code in Listing 2-14 to do this and see the results in Figure 2-7.

***Listing 2-14.*** The Bag-of-Words Approach

```
def tokenize(sent1):
    return word_tokenize(sent1)

df = pd.read_csv('airline_dataset.csv',low_memory=False,encoding = 'latin1')
df["line"]=df["line"].str.lower().str.lstrip().str.rstrip()

###tokenize all rows
df["token_words"] = df["line"].apply(tokenize)

df.head()
```

| Index | line | class | token_words |
|---|---|---|---|
| 0 | when can i web check-in? | check in | [when, can, i, web, check-in, ?] |
| 1 | want to check in | check in | [want, to, check, in] |
| 2 | please check me in | check in | [please, check, me, in] |
| 3 | check in | check in | [check, in] |

***Figure 2-7.*** *Bag-of-words output*

The data in Figure 2-7 is the output with tokens generated for all the documents in the corpus. Now see Listing 2-15, Listing 2-16, and Figure 2-8.

*Listing 2-15.*

```
###get unique list of words
list_all = []
for i in df["token_words"]:
    list_all = list_all + i
list_words = list(set(list_all))
list_words[0:5]
```

```
['check', 'do', '20th', 'hello', 'monday']
```

```
###initialize array for rows=length of dataframe and columns = unique words.
arr1 = np.zeros([len(df),len(list_words)])
```

```
###Loop through the dataframe and mark presence of words in a sentence as 1
for i,row in df.iterrows():
    token_words = row["token_words"]
    for j in token_words:
        if (j in list_words):
            k = list_words.index(j)
            arr1[i][k] = 1
```

```
###Check if we got it right
get_non_zero = np.where(arr1[0]==1)
list_index = list(get_non_zero[0])
[list_words[i] for i in list_index]
```

```
['can', 'when', 'web', 'i']
```

| | dbgirwlkgg | goodevening | forgot | see | abcddef | sure | goodafternoon | late | night | carry | ... | return | baggange | for | loggin | me | u |
|---|---|---|---|---|---|---|---|---|---|---|---|---|---|---|---|---|---|
| 0 | 0.0 | 0.0 | 0.0 | 0.0 | 0.0 | 0.0 | 0.0 | 0.0 | 0.0 | 0.0 | ... | 0.0 | 0.0 | 0.0 | 0.0 | 0.0 | 0.0 |
| 1 | 0.0 | 0.0 | 0.0 | 0.0 | 0.0 | 0.0 | 0.0 | 0.0 | 0.0 | 0.0 | ... | 0.0 | 0.0 | 0.0 | 0.0 | 0.0 | 0.0 |
| 2 | 0.0 | 0.0 | 0.0 | 0.0 | 0.0 | 0.0 | 0.0 | 0.0 | 0.0 | 0.0 | ... | 0.0 | 0.0 | 0.0 | 0.0 | 1.0 | 0.0 |
| 3 | 0.0 | 0.0 | 0.0 | 0.0 | 0.0 | 0.0 | 0.0 | 0.0 | 0.0 | 0.0 | ... | 0.0 | 0.0 | 0.0 | 0.0 | 0.0 | 0.0 |
| 4 | 0.0 | 0.0 | 0.0 | 0.0 | 0.0 | 0.0 | 0.0 | 0.0 | 0.0 | 0.0 | ... | 0.0 | 0.0 | 0.0 | 0.0 | 0.0 | 0.0 |

5 rows × 232 columns

*Figure 2-8.* *The term-document matrix*

***Listing 2-16.***

```
df_mat = pd.DataFrame(arr1)
df_mat.columns = list_words

df_mat.head()
```

Figure 2-8 is a quick visualization of the matrix you created. This is called the **term-document** matrix. Remember that you filled the cells with 0 or 1 representing the presence or absence of a word in a sentence. You can also mark the cells as the frequency of words in the sentence. Also, in the above example, note that you generated 232 features for a corpus size of 398 documents. Hence, you see plenty of zeros in the above matrix. One way to make the matrix dense and make it suitable for a machine learning model is to simply drop words that don't add any meaning to the sentences. In an earlier exercise, you learned about stop words (and, not, the, a). Dropping these words from the features list will reduce the size of the matrix.

Another way to reduce the importance of redundant features is by an inverted weighing of the values of frequent words in the corpus. So when you fill the values in the cells, you want to "punish" the words that are present in all documents. You do this mathematically using the **TF-IDF** formula. TF-IDF computes term frequency and inverse document frequency. This statistic for a given word indicates how important it is as compared to other words in the corpus. The word that appears in fewer documents will have a higher TF-IDF score and the word that appears in all documents will have a lower TF-IDF score.

1. TF (term frequency) in this formula represents the number of times a word (token or feature) appears in a document normalized by the number of words in the document.

    **TF(t) = (Number of times term t appears in a document)/ (Total number of terms in the document).**

2. IDF (inverse document frequency) represents the number of times the given word occurs in a corpus.

    **IDF(t) = log_e(Total number of documents/Number of documents with term t in it).**

3. The final value is obtained by multiplying TF and IDF.

Let's consider a corpus of 1000 documents. Assume that the frequencies of the words "this" and "language" in the corpus are respectively 500 and 50. This is to say the word "this" appears in 500 documents and the word "language" appears in 50 documents. The IDF for "this" is log(1000/500), which is 0.3010299957, and the IDF for "language" is log (1000/50), which is 1.301029996

Consider the two documents in Table 2-5 and calculate the TF-IDF for the words "this" and "language" in them.

***Table 2-5.***  *Calculating the TF-IDF*

| Sentence | TF-this | TF-language | IDF-this | IDF-language | TF-IDF this | TF-IDF language |
|---|---|---|---|---|---|---|
| this is a great piece written. this will be remembered | 0.2 | 0 | 0.301 | 1.301 | 0.060 | 0.000 |
| language models are very popular | 0 | 0.2 | 0.301 | 1.301 | 0.000 | 0.260 |

So going back to your use case, you will compute the TF-IDF for every feature in every document and fill the cell values with it. Having learned about the term-document matrix, Listing 2-17 shows an example for generating the same matrix using the vectorizer method from the scikit-learn library in Python.

***Listing 2-17.***  The Vectorizer Method

```
from sklearn.feature_extraction.text import TfidfVectorizer

tfidf_vectorizer = TfidfVectorizer(min_df=0.0,analyzer=u'word',ngram_
range=(1, 1),stop_words=None)
tfidf_matrix = tfidf_vectorizer.fit_transform(df["line"])
tf1= tfidf_matrix.todense()

tfidf_vectorizer.vocabulary_

{u'03': 0,
 u'12th': 1,
 u'2017': 2,
```

```
u'20th': 3,
u'24th': 4,
u'26th': 5,
u'35': 6,
u'65321': 7,
u'abcddef': 8,
u'abel': 9,
u'able': 10,
u'access': 11,
u'account': 12,
u'afternoon': 13,
u'air': 14,..}
len(tfidf_vectorizer.vocabulary_),tf1.shape

(223, (398L, 223L))

tf1[0:10]

matrix([[0., 0., 0., ..., 0., 0., 0.],
        [0., 0., 0., ..., 0., 0., 0.],
        [0., 0., 0., ..., 0., 0., 0.],
        ...,
        [0., 0., 0., ..., 0., 0., 0.],
        [0., 0., 0., ..., 0., 0., 0.],
        [0., 0., 0., ..., 0., 0., 0.]])
```

The TfidfVectorizer object has few important parameters to reduce the sparsity of the matrix (number of columns):

- mindf is the minimum number of documents in which a word should be present.

- Stop_words is an option to provide stop words or not.

The analyzer can be a word or character depending on the type of token you want. You want to have words for the problem at hand.

The options for ngram_range provide any length of ngrams (bigrams, trigrams, etc). You can have tokens range from bigrams to trigrams, for instance, by using ngram_range(2,3). Once you get a vectorizer object, you fit it on the corpus and get a sparse matrix. The sparse matrix can then be converted to dense. The command tfidf_vectorizer.vocabulary lists the set of tokens done by the vectorizer. As you can see, there are a bunch of numbers and dates which may not add much value to the analysis but if they are normalized into entities like dates, months, etc. can be of a lot of value. You will learn about data normalization in the next section.

## Data Normalization

Data normalization typically includes techniques like generalizing dates, times, and amounts using regex. For instance, you can group all date formats into a word like "Dates." Similarly, all amounts paid or received can be replaced with the word "Money." You can also relabel similar words. For example, in your case, "Baggage" and "Luggage" mean the same thing and hence they could be replaced into a single word. Normalization helps with two things: making the data dense and preparing the data for any unseen variations. For example, take the sentence "My flight is on Sunday and I want to check in." If you build a model considering the previous sentence, then when a new sentence is presented that says, "My flight is on Friday and I want to check in," the model may not perform well. Hence you replace any such words (in this case, days of week) with possibly a common name (in this case, "Dayofweek") to generalize the model and make it more robust.

## Replacing Certain Patterns

The following are a few examples of how regular expressions can help in preprocessing sentences. If you are new to regular expressions, I suggest you take a quick tutorial on regular expressions to get yourself familiar with the syntaxes. See Listing 2-18.

***Listing 2-18.*** Regular Expressions

```
str1 = "I want to be there on 19th"
re.sub("[0-9]+th","datepp",str1)

'I want to be there on datepp'

str1 = "I want to be there on 23-05-18"
str2 = "I want to be there on 23-05"
```

```
import re
print (re.sub("[0-9]+[\/-]+[0-9]+[\/-]*[0-9]*","datepp",str1))
print (re.sub("[0-9]+[\/-]+[0-9]+[\/-]*[0-9]*","datepp",str2))
I want to be there on datepp
I want to be there on datepp
```

Now let's apply the regular expression preprocessing to the corpus. See Listing 2-19.

***Listing 2-19.*** Applying the Regular Expression Preprocessing

```
df["line1"]= df["line"].str.replace('[0-9]+th','datepp')
df["line1"]=df["line1"].str.replace('[0-9]+[\/-]+[0-9]+[\/-]*[0-9]*','datepp')
df["line1"] = df["line1"].str.replace('[0-9]+','digitpp')
df["line1"]= df["line1"].str.replace('[^A-Za-z]+',' ')
```

You have replaced dates and digits with common words. You also cleaned the text with the last statement where you replace anything that is not an English character with a space. Listing 2-20 and Figure 2-9 show an example of word replacements with a common group name. Here you replace common similar meaning words with a common name. I created a file that maps similar words to a group. You will now import and have a look at the file.

***Listing 2-20.***  Using Common Words

```
pp=pd.read_csv('preprocess.csv',low_memory=False,encoding = 'latin1')
pp.head()
```

| | word | class |
|---|---|---|
| 0 | luggage | baggage |
| 1 | bags | baggage |
| 2 | checkin | check in |
| 3 | chckin | check in |
| 4 | check in | check in |

***Figure 2-9.***  *Mapping similar words to a group*

The next step is to replace the words to the corresponding class name. See Listing 2-21 and Figure 2-10.

***Listing 2-21.***

```
def preprocess(l1):
    l2=l1
    for word in pp["word"]:
        if (l1.find(word)>=0):
            newcl = list(pp.loc[pp.word.str.contains(word),"class"])[0]
            l2 = l1.replace(word,newcl)
            break;

    return l2
df["line1"] = df["line1"].apply(preprocess)
df.tail()
```

| | line | class | line1 |
|---|---|---|---|
| 393 | allowance in baggange | baggage | allowance in baggange |
| 394 | allowance in baggage | baggage | allowance in baggage |
| 395 | baggage allowance | baggage | baggage allowance |
| 396 | free baggage | baggage | free baggage |
| 397 | free luggage | baggage | free baggage |

***Figure 2-10.***

You are now set to run a machine learning algorithm on your dataset. For you to know the performance of the dataset on train and test, you need to slice the dataset into train and test. Since you are dealing with text classes, you must do multi-class classification. This means you will have to split the dataset into multiple classes so that each class gets enough representation in train and test. See Listing 2-22.

***Listing 2-22.*** Splitting the Dataset into Multiple Classes

```
from sklearn.model_selection import StratifiedShuffleSplit
sss = StratifiedShuffleSplit(test_size=0.1,random_state=42,n_splits=1)

for train_index, test_index in sss.split(df, tgt):
    x_train, x_test = df[df.index.isin(train_index)], df[df.index.
    isin(test_index)]
    y_train, y_test = df.loc[df.index.isin(train_index),"class"],df.loc[df.
    index.isin(test_index),"class"]
```

You print out the shapes and check once. See Listing 2-23.

***Listing 2-23.*** Printing Shapes

```
x_train.shape,y_train.shape,x_test.shape,x_test.shape
((358, 3), (358,), (40, 3), (40, 3))
```

You can see that `train` and `test` have been split into 358 and 40 rows each.

You now apply the TF-IDF vectorizer on the `train` and `test` split you have. Remember to do this step after the `train` and `test` split. If this step is done before splitting, chances are you are overestimating your accuracies. See Listing 2-24.

***Listing 2-24.*** Applying the TF-IDF Vectorizer

```
tfidf_vectorizer = TfidfVectorizer(min_df=0.0001,analyzer=u'word',ngram_
range=(1, 3),stop_words='english')
tfidf_matrix_tr = tfidf_vectorizer.fit_transform(x_train["line1"])

tfidf_matrix_te = tfidf_vectorizer.transform(x_test["line1"])

x_train2= tfidf_matrix_tr.todense()
x_test2 = tfidf_matrix_te.todense()
x_train2.shape,y_train.shape,x_test2.shape,y_test.shape
((358, 269), (358,), (40, 269), (40,))
```

You can see from the above matrix that `train` and `test` have an equal number of columns. You have a total of 269 features generated from the sentences. You are now ready to apply your machine learning algorithms with word features as independent variables and classes as the dependent variables. Before you apply the machine learning algorithm, you will do a feature selection step and reduce the features to 40% of the original features. See Listing 2-25.

***Listing 2-25.***  The Feature Selection Step

```
from sklearn.feature_selection import SelectPercentile, f_classif
selector = SelectPercentile(f_classif, percentile=40)
selector.fit(x_train2, y_train)
x_train3 = selector.fit_transform(x_train2, y_train)
x_test3 = selector.transform(x_test2)
x_train3.shape,x_test3.shape
((358, 107), (40, 107))
```

You are ready to run your machine learning algorithm on the train dataset and test it on x_test. See Listing 2-26 and Listing 2-27.

***Listing 2-26.***

```
clf_log = LogisticRegression()
clf_log.fit(x_train3,y_train)
```

***Listing 2-27.***

```
pred=clf_log.predict(x_test3)
print (accuracy_score(y_test, pred))
0.875
```

A quick model has a higher accuracy as compared to the rules that you built earlier. This result is on an unseen test dataset unlike the rules that were tested on the same dataset and hence this result is more robust. You will now test on some of the other metrics of importance. In a multiclass problem, class imbalance is a common occurrence where the distribution of categories could be biased towards a few of the top categories. Hence over and above the overall accuracy you want to test individual accuracies for different categories. This can be done using a confusion matrix, precision, recall, and F1 measures. You can learn more about these scores from the article titled "Accuracy, Precision, Recall & F1 Score: Interpretation of Performance Measures" at https://blog.exsilio.com/all/accuracy-precision-recall-f1-score-interpretation-of-performance-measures/ as well as from Scikit-Learn's pages on confusion matrix (https://scikit-learn.org/stable/modules/generated/sklearn.metrics.confusion_matrix.html) and F1 scores (https://scikit-learn.org/stable/modules/generated/sklearn.metrics.f1_score.html#sklearn.metrics.f1_score). Now see Listings 2-28 and 2-29 and Figure 2-11.

*Listing 2-28.*

```
from sklearn.metrics import f1_score
f1_score(y_test, pred, average='macro')
0.8060606060606059
```

*Listing 2-29.*

```
from sklearn.metrics import confusion_matrix
confusion_matrix(y_test, pred, labels=None, sample_weight=None)
                    array([[ 7,  0,  0,  0,  0,  1,  0],
                           [ 0,  2,  0,  0,  0,  0,  0],
                           [ 0,  0,  5,  0,  0,  1,  0],
                           [ 0,  0,  0,  4,  0,  0,  0],
                           [ 0,  0,  0,  0, 10,  0,  0],
```

*Figure 2-11.* *A confusion matrix*

You can see that the confusion matrix has strong diagonal matrix and is indicative of a good model with few classification errors.

# Identifying Issue Lines

In the example above, the data provided were single-line sentences. However, in a conversation there is a whole corpus of dialogue. Tagging the entire dialog corpus to an intent is difficult. The signal-to-noise ratio is pretty low, so in general you try to identify portions of dialog first that could contain the issue lines and then mine only those sentences for the intent. This improves the signal-to-noise ratio significantly. You can also build another supervised learning model to locate which of the lines in the dialog contain the issues of the customers. The customer issue typically is in the first few lines of the "chat" or it would follow a question like "how may I help you". See Table 2-6.

***Table 2-6.*** *Issue Lines*

| System | Thank you for choosing Best Telco. A representative will be with you shortly. |
|---|---|
| System | You are now chatting with Max. |
| Customer | Himax |
| Agent | Thank you for contacting Best Telco. My name is Max. |
| Agent | Hello, I see that I'm chatting with Mrs. Sara and you've provided XXXXXXX as the number associated with your account. Is that correct? |
| Customer | yes. that is correct. I just received email saying my service price was going up to$70/monthbut on your web site, it's only$30. What's the deal with that? |

As you can see in the above example, the line in which the customer states the problem is the second line by the customer and directly follows the greeting of the user. In emails, it may be the starting line of a thread by the customer. So locating and looking for issue lines yields better results.

Having gone through an in-depth look at intent mining as well as topic modelling both by supervised and by unsupervised mechanisms, you will now look at some classic analysis in customer service where these intents form the backbone. It is like getting the base of a pizza right and then tossing on different toppings. Once you have the intent of a customer service conversation, you can do the following analysis.

# Top Customer Queries

Frequency distribution of queries is a quick insight into trending topics and why customers reach out in general. For the examples here, you will use the `airline` dataset. You will create the dataframe with the frequency distribution of the labels. Please note the additional column named `CSAT` in the `airline` dataset. This is the CSAT rating provided by the users. For charting, you will use the matlobplotlib library. See Listing 2-30 and Figure 2-12.

*Listing 2-30.*

```
import pandas as pd
import numpy as np
import matplotlib.pyplot as plt
df = pd.read_csv('airline_dataset_csat.csv',low_memory=False,encoding = 'latin1')
df.head()
```

|   | line | class | csat_fnl |
|---|------|-------|----------|
| 0 | When can I web check-in? | check in | 5 |
| 1 | want to check in | check in | 5 |
| 2 | please check me in | check in | 5 |
| 3 | check in | check in | 5 |
| 4 | my flight is tomm can I check in | check in | 5 |

*Figure 2-12.*

Now you will get the frequency distribution and plot the top queries with the absolute values. See Listing 2-31 and Figure 2-13.

*Listing 2-31.*

```
freq1= pd.DataFrame(df["class"].value_counts()).reset_index()
freq1.columns = ["class","counts"]
freq1["percent_count"] = freq1["counts"]/freq1["counts"].sum()
freq1
```

| | class | counts | percent_count |
|---|---|---|---|
| 0 | login | 105 | 0.263819 |
| 1 | other | 79 | 0.198492 |
| 2 | baggage | 76 | 0.190955 |
| 3 | check in | 61 | 0.153266 |
| 4 | greetings | 45 | 0.113065 |
| 5 | thanks | 16 | 0.040201 |
| 6 | cancel | 16 | 0.040201 |

*Figure 2-13.*

You now plot the values using pyplot in matlab. `pyplot.bar` takes X and Y values to build the bar chart. See Listing 2-32 and Figure 2-14.

*Listing 2-32.* Plotting the Values

```
x_axis_lab = list(freq1["class"])
y_axis_lab = list(freq1["counts"])
y_per_axis_lab = list(freq1["percent_count"])

import matplotlib.pyplot as plt
plt.bar(x_axis_lab,y_axis_lab)
plt.ylabel('Counts')
plt.xlabel('Categories')
plt.show()
```

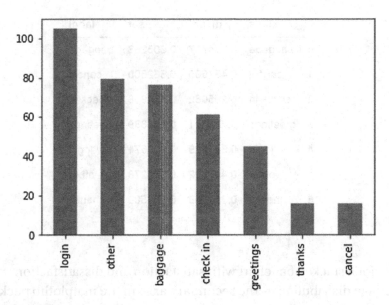

*Figure 2-14.*  *Plotting the values*

# Top CSAT Drivers

Customer satisfaction is a survey that users answer at the end of a customer service interaction, generally on a scale of 1 to 5. Plotting CSAT responses by topics mined gives an idea of which intents customers are satisfied or dissatisfied with. In the `airline` dataset example, you classify customers as satisfied if they give a rating of 4 and 5. A score of 1 to 3 means the customers are considered to be dissatisfied. You classify the `csat_fnl` score based on this logic. You first cross-tab the intent and satisfaction class (`satisfied`, `dissatisfied`). See Listing 2-33 and Figure 2-15.

*Listing 2-33.*

```
df["sat"]=1
df.loc[df.csat_fnl<=3,"sat"]=0
ct = pd.crosstab(df["class"],df["sat"],normalize='index')

ct = ct.reset_index()
ct_cols = ct.columns
ct_cols1 = [ct_cols[0],"dissat","sat"]
ct.columns = ct_cols1
ct["label"] = ct["class"]
ct
```

| | class | dissat | sat | label |
|---|---|---|---|---|
| 0 | baggage | 0.394737 | 0.605263 | baggage |
| 1 | cancel | 0.437500 | 0.562500 | cancel |
| 2 | check in | 0.295082 | 0.704918 | check in |
| 3 | greetings | 0.311111 | 0.688889 | greetings |
| 4 | login | 0.371429 | 0.628571 | login |
| 5 | other | 0.417722 | 0.582278 | other |
| 6 | thanks | 0.312500 | 0.687500 | thanks |

***Figure 2-15.***

You now plot a stacked bar chart with satisfaction and dissatisfaction. You use the intent percentage distribution as the secondary axis. In the matplotlib package, you create subplots in order to show two y-axes. First, you create the stacked bar for satisfied/dissatisfied using the parameter bottom. The dissat variable stacks on top of the csat variable. See Listing 2-34.

***Listing 2-34.***

```
ct_sat = ct["sat"]
ct_dissat = ct["dissat"]
ind = ct["label"]

fig,ax=plt.subplots()
p1 = ax.bar(ind, ct_sat)
p2 = ax.bar(ind, ct_dissat,
            bottom=ct_sat)
ax.set_xlabel('Categories')
ax.set_ylabel('Sat/Dissat percentage')

ax2=ax.twinx()
# make a plot with different y-axis using second axis object
ax2.plot(ind, y_per_axis_lab,color="blue",marker="o")

ax2.set_ylabel('intent percetage')
plt.show()
```

The ax.twinx() function helps you to create the secondary y-axis. The variable y_per_axis_lab in the ax2.plot function contains the percentage distribution of intents as see in. The final chart is shown in Figure 2-16.

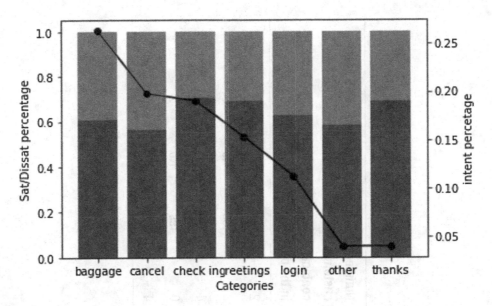

***Figure 2-16.***

Similar charts can be plotted with CSAT, overall handle time of chats, resolution rates, and so on, and the actionability of these analyses can then be derived.

# Top NPS Drivers

A quick word on NPS before you move further in this section. Top NPS Drivers is the net promoter score. At the end of an interaction, the customer is asked how likely they are to recommend to their friends/relatives a product/service on a scale of 1 to 10. Customers who provide an input from 8-10 are considered promoters, 1-4 are detractors, and 5-7 are neutral. The NPS formula is

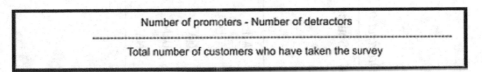

The graph you saw in the last section was about how user issues or intents correlate with CSAT. The same could be plotted with NPS as well. In some cases, you may want to go deeper and know the factors that are causing a difference in CSAT or NPS. NPS, for instance, could be a function of intent, resolution provided, courtesy or politeness of the agent, product attributes, user variables, and more. Table 2-7 shows a sample dataset of NPS.

*Table 2-7. Sample Dataset of NPS*

| User_id | Intent | AHT | Source | Destination | Budget | Family size | Type of customer | Number of customer contacts in the past | Resolved | Referred to call | NPS score |
|---|---|---|---|---|---|---|---|---|---|---|---|
| 50307 | Booking | 16 | ATL | FLL | 600 | 4 | Gold | 2 | Yes | No | 8 |
| 40697 | Checkin | 17 | ATL | LGA | NA | NA | New | 3 | Yes | No | 9 |
| 23042 | Login | 16 | NA | | NA | NA | New | 4 | Yes | No | 9 |
| 35857 | Baggage | 11 | DEN | LAX | NA | NA | Silver | 0 | Yes | No | 2 |
| 24394 | Login | 20 | NA | | NA | NA | Silver | 4 | No | Yes | 5 |

The idea here is to understand the top drivers of NPS. You do a regression model to understand the top drivers of NPS and hence a rank-ordered priority of action of items. Mathematically, it can be represented as

$$NPS = f\left(user\ attributes, transaction\ attributes, interaction\ attributes, agent\ attributes\right)$$

I will discuss each of the attributes in detail below:

- **User attributes**: These are user characteristics like user membership, age, average spend, lifetime value, and so on. These attributes are available as structured data in some databases. If they are not available, they can be either extracted from current interactions or historic interactions of the user. It is also advisable to log these variables because they can provide a better understanding of the problem at hand. For instance, if you find that the NPS is low for new customers, then there is a major problem at hand as the growth of organization could be impacted. In order to extract some of the attributes from the interaction data, you can follow a rules-based approach.

- **Transaction attributes:** These are the characteristics that define the specific transaction behind the intent. So if it's a booking enquiry, this means the source, destination, budget, and number of people. Or in case of a cancellation, it could be route, mode of payment, or cancellation fee. These attributes can critically influence resolution rates and hence can influence NPS.

- **Interaction attributes:** These are the characteristics of the current interaction, such as number of customer lines, number of agent lines, handle time of the chat (end time of the chat minus start time of the chat), average response time of the customer, average response time of the agent (total response time of the agent divided by number of responses), overall sentiment of the chat, number of resolutions provided, was the issue resolved or not, and escalations (to a supervisor or other departments) if any.

- **Agent attributes:** These are the characteristics of the support agent who handles the interaction such as if they are courteous and polite in the conversation and if they are knowledgeable about the product. There are a couple of ways to extract these attributes: by a quality specialist who can rate these interactions in different dimensions or by extracting from text. Since its not scalable to rate these interactions by a quality specialist, text mining approach is the best way to solve this problem.

Let's see some examples from Customer Thermometer (`www.customerthermometer.com/customer-service/excellent-customer-service-phrases/`):

1. I am sorry to hear that you feel this way [Mr. X/Ms. X].

2. Your feedback is enormously valuable to us so we greatly appreciate you taking the time to call [Mr. X/Ms. X].

3. I'd like to call you back to give you an update. When would be the best time to reach you?

4. I completely understand how you feel [Mr. X/Ms. X].

5. I fully appreciate the inconvenience this has caused you [Mr. X/Ms. X].

6. Thank you for your understanding [Mr. X/Ms. X]. We are doing everything we can to resolve your problem quickly.

The above sentences express courtesy as opposed to sentences like "We won't be able to do anything here," "Sorry it can't be helped," and "This is how we work." Commonly variables that define the service quality are used to explain customer support agent behavior. (For more information, see the article titled "Applying The Technology Acceptance And Service Quality Models To Live Customer Support Chat For E-Commerce Websites" at `www.researchgate.net/publication/289170615_Applying_The_Technology_Acceptance_And_Service_Quality_Models_To_Live_Customer_Support_Chat_For_E-Commerce_Websites`). Also see Table 2-8.

***Table 2-8.*** *Dimensions and Customer Support*

| Dimension | How it applies to customer support |
|---|---|
| Reliability | Customer support agents being reliable with responses |
| Responsiveness | How fast questions get answered |
| Assurance | Can they assure if the issues will be answered |
| Empathy | Can they show empathy towards a customer's concerns |
| Tangibles | Widgets used, the look and feel of the chat window itself |

The paper titled "A human capital predictive model for agent performance in contact centres" (www.researchgate.net/publication/262612058_A_human_capital_ predictive_model_for_agent_performance_in_contact_centres) also describes the different dimensions to measure agent performance. Added to the above list could be agent's knowledge/competence, their motivation to solve the problem, and so on. You can take a rules-based approach to solve these problems or use online lexicons for mining some of the behavior attributes. As an example, see "Empathy Statements for Customer Service Representatives" (www.customerservicemanager.com/empathy- statements-for-customer-service-representatives/) for examples of empathy statements. Similarly, with domain experts, you could arrive at lexicons for other attributes as well. You can follow a similar approach as the heuristic process described earlier to classify each sentence of the agent in a conversation and then arrive at an overall score to model NPS. Table 2-9 shows a small example of agent behavior tagging. When you try to mine the text to score the dimensions, you only score it on agent lines and not on the customer lines. In incase of the text corpus being emails you can locate the sender tags and mine only those lines.

**Table 2-9.**  *Agent Behavior Tagging*

| Speaker | Sentence | Empathy | Competency | Assurance |
|---|---|---|---|---|
| Diane | Thanks for contacting Bright Smiles Dentistry. My name is Diane. Hi, how may I help you? | 0 | 0 | 0 |
| Aaron | I received an email from Groupon for your teeth whitening service for $99. | 0 | 0 | 0 |
| Diane | Thanks for your interest in Bright Smiles. We have gotten amazing response from our Groupon advertisement. Could we schedule an appointment? | 0 | 0 | 0 |
| Aaron | How long does the procedure usually take? | 0 | 0 | 0 |
| Diane | The $99 teeth whitening special is for our basic service. Generally, these appointments take about 20 minutes. | 0 | 1 | 0 |
| Aaron | Do you take dental insurance? | 0 | 0 | 0 |
| Diane | Bright Smiles is a cosmetic dental center and unfortunately we do not take any insurance. Payment can be made either via cash, Visa, MasterCard. | 1 | 0 | 0 |
| Aaron | Does the procedure hurt? | 0 | 0 | 0 |
| Diane | Absolutely no. Let me assure you it is fast and painless. | 0 | 0 | 1 |
| Aaron | All right, what the heck? I'll schedule an appointment. Do you have anything this afternoon? | 0 | 0 | 0 |
| Diane | Actually, we do. Could you get here by 3:30 PM? | 0 | 1 | 0 |
| Aaron | Yes. Diane Let me take your name and phone number. | 0 | 0 | 0 |
| Aaron | Aaron Davis and my number is 610-265-1715. | 0 | 0 | 0 |

*(continued)*

***Table 2-9.*** (*continued*)

| Speaker | Sentence | Empathy | Competency | Assurance |
|---------|----------|---------|------------|-----------|
| Diane | Do you need directions to our offices? | 0 | 0 | 0 |
| Aaron | No. I know exactly where you are; Right next to the Baskin Robins. | 0 | 0 | 0 |
| Diane | Great; you will need to print out your Groupon, or have the receipt on the phone. | 0 | 1 | 0 |
| Aaron | O.K. I will purchase the package now, and print that out. | 0 | 0 | 0 |
| Diane | Do you have any other questions? Or is there anything else that I can do for you? | 0 | 0 | 1 |
| Aaron | Not that I can think of. Thanks for your help. | 0 | 0 | 0 |
| Diane | Thank you for choosing Bright Smiles Dentistry. | 0 | 0 | 0 |

You can now sum up the scores for the above interaction on the three dimensions of empathy, competency, and assurance. See Figure 2-17.

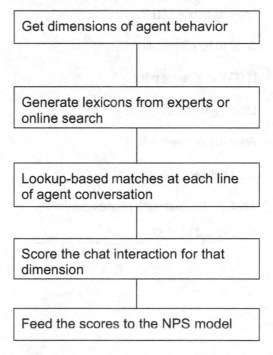

***Figure 2-17.*** *NPS scoring*

If you have tagged a corpus of sentences to different agent performance dimensions, then you can also build a supervised algorithm to classify the sentences to different categories.

# Insights into Sales Chats

The examples and the analysis you have seen so far are for service use case where the customer service agent is answering queries and not selling something. In sales chats, you want to understand two important things: what product the customer came for and why they purchased it (or more importantly, why they didn't purchase it).

# Top Products for Sales Chats

You can use named entity recognition techniques, which you will see in subsequent chapters for product extraction, or you can use a lookup-based method for extracting product mentions. Table 2-10 shows some example sentences.

***Table 2-10.*** *Example Sentences*

| |
|---|
| Looking to buy LED TV |
| Want to buy Samsung LED |
| LED TV is at what price |
| Looking to buy fridge |
| Refrigerator need help |
| need help with fridge |
| what are the samsung phones available |
| is S340 out of stock |

Products can have various variations. For instance, a refrigerator can be called a fridge or by a brand name like Samsung or a model name like 687L or sometimes a brand name with capacity like Samsung 200 L. Extracting products for retail or ecommerce is very complex given the number of possibilities of products out there and the possible ways of referencing them. You will see this in detail in Chapter 3. For some of the other domains, like telco or BFSI, the variation in products is not large and hence they can be handled with simple lookups. Figure 2-18 shows a chart you could analyze and get insights based on conversion rates and volume. It shows top products with the conversion rates of the products.

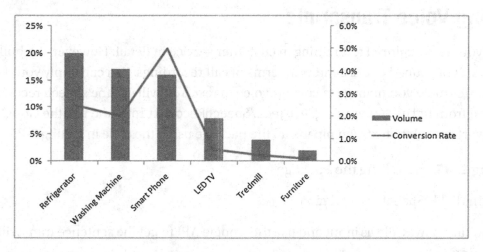

*Figure 2-18. Top products and conversion rates*

# Reasons for Non-Purchase

You can also mine the chats for objections raised by customers about buying the product. Normally a good classifier trained on objections can do well. Here, locating the issue line is the key. Not every line in the chat will have an objection from the customer. Sometimes agents are also prompted to ask leading questions to customers to state their objections. The reasons for non-purchase are also collected as part of the final agent dispositions. The top reasons why customers generally do not buy are price, incompatibility of product, installation, return policy, shipping cost, better deal with a competitor, and wanting to consult others before making a purchase decision. These high-level reasons are similar across different companies, so you can use the same predictive model (or even heuristic rules) for different organizations.

# Survey Comments Analysis

Surveys are collected at the end of a chat to understand the customer experience. This dataset proves very rich in trying to understand the customer's perception about the agent or product or even reasons for non-purchase. A set of heuristics or a supervised model is applied to mine survey comments. This corpus is also a good indicator of emotions or sentiments of the user. Chat conversations generally tend to be neutral whereas the survey comments can clearly indicate the sentiment of a customer. You will look at sentiment mining in detail in the next chapter.

# Mining Voice Transcripts

So far you have explored text mining in customer service in detail. However, the bulk of data is still produced in call centers in terms of call recordings. You can apply your text techniques once you manage to convert voice to text. You will use the speech recognition package from `https://pypi.org/project/SpeechRecognition/` and use the Google API to convert speech to text. You pip install the package using the code in Listing 2-35.

***Listing 2-35.*** Installing the Package

```
!pip install SpeechRecognition
```

You take a .wav file as input and use the Google API to get the sentence transcribed. See Listing 2-36.

***Listing 2-36.*** Transcribing Sentences

```
import speech_recognition as sr
r = sr.Recognizer()

def speech_to_text(af,show_all):
    #playsound(af)

    with sr.AudioFile(af) as source:
        #reads the audio file. Here we use record instead of
        #listen

        audio = r.record(source)

        abc = r.recognize_google(audio,show_all=show_all)
    return abc
```

The function takes two arguments. One is the file itself. The other is the speech recognition's flag called show_all. If this is False, then the Google API returns the most probable speech-to-text recognition. The output is shown in Listing 2-37.

***Listing 2-37.***

```
af1 = "sample_audio.wav"
speech_to_text(af1,False)
'I am calling for reversal of enquiry and my account number is 911'
```

Setting show_all to False, the Google API returns all of the recognition with the confidence level as a dictionary. This can be useful if you want to pick up the most probable outputs. Now see Listing 2-38.

***Listing 2-38.***

```
speech_to_text(af1,True)
{'alternative': [{'transcript': 'I am calling for reversal of
enquiry and my account number is 911',
   'confidence': 0.89367926},
  {'transcript': 'I am calling for reversal of enquiry in my account
  number is 911'},
  {'transcript': 'I am calling for reversal of enquiry and my account
  number is 9 11'},
  {'transcript': "I'm calling for reversal of enquiry and my account
  number is 911"},
  {'transcript': 'I am calling for the reversal of enquiry and my account
  number is 911'}],
 'final': True}
```

There are other APIs that the package supports like Bing, IBM, and Spinx. Go to https://pypi.org/project/SpeechRecognition/.

You can also build your own speech recognition tool using a neural network model ground-up. This is outside the scope of this book, but I will summarize the high-level steps included in speech recognition. There are two essential parts. One is the acoustic model that converts speech to text and the other is a language model that makes the text more meaningful and coherent.

# Acoustic Model

The first step is to get the data from the audio files along with the corresponding labels. There are a lot of open source datasets to get this step started. Once you have the files, you convert the .wav file into numbers using digital signal processing techniques. Some standard practices are to convert them using the MFCC or Spectrogram approach.

# Language Model

After you do this, you have an array of numbers and corresponding labels. You can then train a neural network model to classify a given set of arrays to the words it could possibly contain. In the case of scoring a new audio file, the file is first segmented into smaller audio chunks and then each of the chunks is recognized as a word using the speech recognition model. Once you have the set of words recognized, converting them to a meaningful and most probable sentence or a phrase is the job of a language model. There are cases where the acoustic model is trained on detecting a letter of a word. The language model gets the most probable word.

The language model is based on the probability of words occurring together. There can be a pure frequency-based approach or there can be a recurrent neural network-based sequence to sequence the model used to correct the sentences coming out of the acoustic model. Listings 2-39 and 2-40 show an example of building a language model using bigrams and probability distribution. They use a few lines from Wikipedia on natural language processing.

***Listing 2-39.*** Building a Language Model

```
import pandas as pd
from nltk.util import ngrams
import numpy as np
from collections import Counter
import re
```

***Listing 2-40.***

```
inp_text=open("nlp_text.txt","r").read()
###print characters
Inp_text[0:100]
```

'Natural language processing\n\n\nAn automated online assistant providing customer service on a web page'

You do some preprocessing and clean the text so that all words are normalized. You can also remove stop words depending on the use case at hand. See Listing 2-41.

***Listing 2-41.*** Cleaning Up the Text

```
inp_text = inp_text.lower()
inp_text1 = re.sub('[^a-z/s]+',' ',inp_text)
Inp_text1[0:1000]
```

'natural language processing an automated online assistant providing customer service on a web page an example of an application where natural language processing is a major component natural language processing nlp is a subfield of linguistics computer science information engineering and artificial intelligence concerned with the interactions between computers and human natural languages in particular how to program computers to process and analyze large amounts of natural language data challenges in natural language processing frequently involve speech recognition natural language understanding and natural language generation contents history rule based vs statistical nlp major evaluations and tasks syntax semantics discourse speech dialogue see also references further reading history the history of natural language processing nlp generally started in the s although work can be found from earlier periods in alan turing published an article titled computing machinery and intelligence w'

Next step is to get a corpus of bigram words from the sentences. When building a language model, once you provide the word, the model will output the most probable word following the given word. You can increase the n-grams to a higher number for better precision. See Listing 2-42.

***Listing 2-42.***

```
n_grams = ngrams(inp_text1.split(), 2)
l1 = []
for grams in n_grams :
    l1.append((grams[0].lower(),grams[1].lower()))
```

The bigrams are now organized as a dataframe with lead and lag words. So given lead word, you can get the set of probable lag words. See Listings 2-43 and 2-44.

***Listing 2-43.***

```
df0 = pd.DataFrame(l1)
df0.columns = ["lead","lag"]

lead_all = df0["lead"].unique()
lag_all = df0["lag"].unique()
```

***Listing 2-44.***

```
lead_dict = {}
for i in lead_all:
    matches = df0.loc[df0.lead==i,"lag"]
    len_mtch = len(matches)
    lag_dict =  dict(Counter(matches))
    for k in lag_dict:
        lag_dict[k] = lag_dict[k]/len_mtch

    lead_dict[i] = lag_dict
```

Let's test this for the word "language." Listing 2-45 shows the most probable words following the word "language."

***Listing 2-45.***  Most Probable Words Test

```
lead_dict["language"]
{'processing': 0.6666666666666666,
 'data': 0.041666666666666664,
 'understanding': 0.041666666666666664,
 'generation': 0.041666666666666664,
 'system': 0.041666666666666664,
 'models': 0.041666666666666664,
 'tasks': 0.041666666666666664,
 'modeling': 0.08333333333333333}
```

This language model is also used in certain other use cases like machine translation or natural language generation. So going back to your example, the output of the acoustic model is corrected using a language model and then you get a voice-to-text output. Once you get the text output, it can be mined and all the insights will be derived.

So far you have seen a good number of use cases of NLP in the customer service industry. There are more nuanced use cases as well, like extracting personas, shifting of sentiments as the conversation proceeds, correlation of different customer service channels like voice and text, and so on. The use cases that have been covered here should be adequate for anyone involved in this area to get started in that area. Customer service is rich with text data, and analyzing different dimensions of this data can provide a significant impact to organizations.

# CHAPTER 3

# NLP in Online Reviews

Reviews are a significant part of online buying cycles today. Though the "vocal minority" is few, the number of users who are impacted by reviews is significantly large. One study found that 63% of users prefer online sites that have reviews. Customers who visit review pages have an astounding 105% more chance of buying from the website (`https://cxl.com/blog/user-generated-reviews/#:~:text=Reevoo%20found%20that%2050%20or,site%20that%20has%20user%20reviews`). Mining these reviews gives insights to both the online service provider as well as the seller who has listed the product. Other than knowing whether the customer is happy or not, we can also know how users feel about each feature in the product. Sometimes reviewers write a lot about their lifestyle and the use case they have found for the product as well. This can provide insights into things like product-market fit or the value proposition for the product. This can later be used in brand communications for the product. We can also find opportunities or gaps in a category and hence get the "voice of the customer" to create a new product or even start a new business.

## Sentiment Analysis

Sentiment is the first and most common attribute extracted from reviews. Sentiment is defined as the feeling or emotion expressed by the user in the given corpus. This is generally expressed as a positive or negative sentiment. But there are other dimensions of a sentiment like anger, delight, frustration, and so on. Yet another dimension can measure the objectivity or subjectivity of the text corpus. Figure 3-1 shows some examples.

65

© Mathangi Sri 2021

M. Sri, *Practical Natural Language Processing with Python*, https://doi.org/10.1007/978-1-4842-6246-7_3

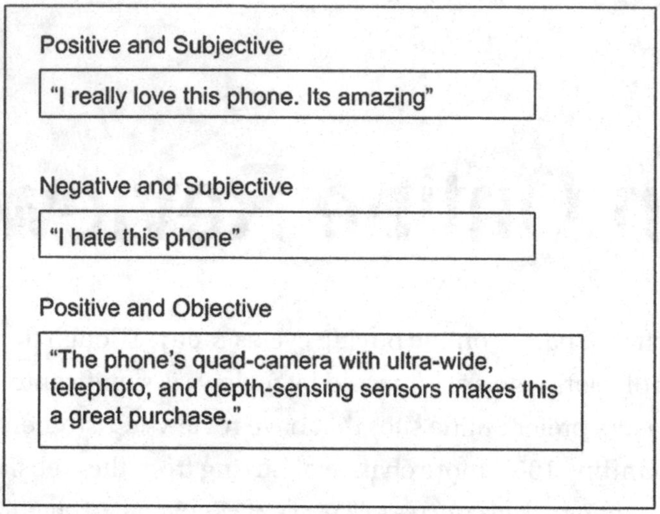

Positive and Subjective

"I really love this phone. Its amazing"

Negative and Subjective

"I hate this phone"

Positive and Objective

"The phone's quad-camera with ultra-wide, telephoto, and depth-sensing sensors makes this a great purchase."

*Figure 3-1.* *Sentiments*

# Emotion Mining

Sentiment analysis is a simplified version of analyzing deeper emotional feelings. Emotions have been well researched in the field of psychology. In certain product review mining use cases, we can go beyond sentiment analysis and into deeper emotion mining. One of the well-known theoretical models of emotions is Plutchik's wheel of emotions (Figure 3-2). The wheel of emotions at the core consists of eight basic emotions: joy, sorrow, anger, fear, trust, disgust, surprise, and anticipation. The intensity reduces as one moves towards the outer wheel and intensifies as one moves towards the core of the wheel. The white area in the wheel is the combination of two emotions. Even if you're not trying to mine the whole wheel, mining the eight basic emotions can reveal a lot about user preferences. For further reader, go to "A Survey on Sentiment and Emotion Analysis for Computational Literary Studies" at https:// arxiv.org/abs/1808.03137.

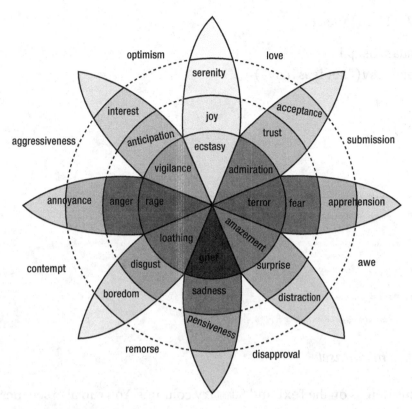

***Figure 3-2.*** *Plutchik's wheel of emotions*

Let's now get into the code and the details of how to do sentiment mining in Python. You will first build a basic version and then I will discuss how complex use cases can be tackled from there. You will be using the Amazon review dataset for this chapter. All packages and their version numbers are listed in Table 3-7 at the end of the chapter.

# Approach 1: Lexicon-Based Approach

The first approach is simple. You will be using a standard lexicon-based approach to classify positive, negative, and neutral sentiments. Lexicons are a list of words that represent a topic. You will use some lexicons available on the Internet for classifying positive, negative, and neutral reviews.

**Dataset:** You will use the Amazon Fine Food reviews dataset for this analysis. You can download it from Kaggle; it's the Survey for Amazon Fine Food Reviews at `www.irjet.net/archives/V6/i4/IRJET-V6I4134.pdf` See Listing 3-1 and Figure 3-3.

***Listing 3-1.*** The Dataset

```
import pandas as pd
t1 = pd.read_csv("Reviews.csv")

t1.shape
(568454, 12)

t1.head()
```

| | Id | ProductId | UserId | ProfileName | HelpfulnessNumerator | HelpfulnessDenominator | Score | Time | Summary | Text |
|---|---|---|---|---|---|---|---|---|---|---|
| 0 | 1 | B001E4KFG0 | A3SGXH7AUHU8GW | delmartian | 1 | 1 | 5 | 1303862400 | Good Quality Dog Food | I have bought several of the Vitality canned d... |
| 1 | 2 | B00813GRG4 | A1D87F6ZCVE5NK | dll pa | 0 | 0 | 1 | 1346976000 | Not as Advertised | Product arrived labeled as Jumbo Salted Peanut... |
| 2 | 3 | B000LQOCH0 | ABXLMWJIXXAIN | Natalia Corres "Natalia Corres" | 1 | 1 | 4 | 1219017600 | "Delight" says it all | This is a confection that has been around a fe... |
| 3 | 4 | B000UA0QIQ | A395BORC6FGVXV | Karl | 3 | 3 | 2 | 1307923200 | Cough Medicine | If you are looking for the secret ingredient i... |
| 4 | 5 | B006K2ZZ7K | A1UQRSCLF8GW1T | Michael D. Bigham "M. Wassir" | 0 | 0 | 5 | 1350777600 | Great taffy | Great taffy at a great price. There was a wid... |

***Figure 3-3.*** *The dataset*

Our focus here is on the Text and Summary columns. You can also see the Score column. You will be using the score to understand and measure the accuracy of your lexicon-based approach. You will be using the lexicons mentioned in the paper called "Mining and summarizing customer reviews" at https://dl.acm.org/doi/10.1145/1014052.1014073. See Listing 3-2.

***Listing 3-2.***

```
pos1=pd.read_csv("positive-words.txt",sep="\t",error_bad_lines=True,
encoding='latin1',header=None)
neg1=pd.read_csv("negative-words.txt",sep="\t",error_bad_lines=True,
encoding='latin1',header=None)

pos1.columns = ["words"]
neg1.columns = ["words"]

pos_set = set(list(pos1["words"]))
neg_set = set(list(neg1["words"]))
```

In the t1 dataset are two columns of text: Text and Summary. You will combine them and process them into a single column. It is this column that will go through the lexicon mining. See Listing 3-3.

***Listing 3-3.*** Combining Columns

```
t1["full_txt"] = t1["Summary"] + " " + t1["Text"]
t1["full_txt"] = t1["full_txt"].str.lower()
t1["sent_len"] = t1["full_txt"].str.count(" ") + 1
```

Listing 3-4 helps you leave out sentences that have missing words.

***Listing 3-4.*** Refining the Dataset

```
t2 = t1[t1.sent_len>=1]
len(t1),len(t2)
(568454, 568427)
```

In order to measure the accuracy of your approach, you will bucket the final rating provided by customer in the Score column into buckets. See Listing 3-5.

***Listing 3-5.***

```
##for meausring accuracy
t2["score_bkt"]="neu"
t2.loc[t2.Score>=4,"score_bkt"] = "pos"
t2.loc[t2.Score<=2,"score_bkt"] = "neg"
```

The simplest approach is to iterate through all of the words in a sentence corpus and hit against the list of lexicons. Since these are longer sentences, you want to normalize the number of positive and negative hits by the number of words in the sentences. After this, a simple comparison between the positive, negative, and neutral scores is done and the sentences are tagged based on whichever scores are higher. In order for you to improve further and come up with additional strategies, you also log the list of words that were present in the sentences as pos_set_list and neg_set_list. See Listing 3-6.

***Listing 3-6.***

```
final_tag_list = []
pos_percent_list = []
neg_percent_list = []
pos_set_list = []
neg_set_list = []

for i,row in t3.iterrows():

    full_txt_set = set(row["full_txt"].split())
    sent_len = len(full_txt_set)

    pos_set1 = (full_txt_set) & (pos_set)
    neg_set1 = (full_txt_set) & (neg_set)

    com_pos = len(pos_set1)
    com_neg = len(neg_set1)

    if(com_pos>0):
        pos_percent = com_pos/sent_len
    else:
        pos_percent = 0

    if(com_neg>0):
        neg_percent = com_neg/sent_len
    else:
        neg_percent =0

    if(pos_percent>0)|(neg_percent>0):
        if(pos_percent>neg_percent):
            final_tag = "pos"
        else:
            final_tag = "neg"
    else:
        final_tag="neu"
```

```
final_tag_list.append(final_tag)
pos_percent_list.append(pos_percent)
neg_percent_list.append(neg_percent)
pos_set_list.append(pos_set1)
neg_set_list.append(neg_set1)
```

Assigning the dataframe of t3 with the list created is done in Listing 3-7.

***Listing 3-7.*** Assigning the List

```
t3["final_tags"] = final_tag_list
t3["pos_percent"] = pos_percent_list
t3["neg_percent"] = neg_percent_list

t3["pos_set"] = pos_set_list
t3["neg_set"] = neg_set_list
```

Now you have classified the sentences into three categories of positive, negative, and neutral. You would like to check if they are aligned with the score_bkt discussed earlier. You are going to measure using the accuracy score, F1 score, and confusion metrics. More details about the metrics can be learned from the article titled "Accuracy, Precision, Recall & F1 Score: Interpretation of Performance Measures" at https://blog.exsilio.com/all/accuracy-precision-recall f1-score-interpretation-of-performance-measures/ as well as from Scikit-Learn's pages on confusion matrix (https://scikit-learn.org/stable/modules/generated/sklearn.metrics.confusion_matrix.html) and F1 score (https://scikit-learn.org/stable/modules/generated/sklearn.metrics.f1_score.html#sklearn.metrics.f1_score). Now see Listing 3-8 and Figure 3-4.

***Listing 3-8.***

```
from sklearn.metrics import accuracy_score
print (accuracy_score(t3["score_bkt"],t3["final_tags"]))
#0.8003448093872596

from sklearn.metrics import f1_score
f1_score(t3["score_bkt"],t3["final_tags"], average='macro')
#0.502500231042986
```

```
rows_name = t3["score_bkt"].unique()

from sklearn.metrics import confusion_matrix
cmat = pd.DataFrame(confusion_matrix(t3["score_bkt"],t3["final_tags"],
labels=rows_name, sample_weight=None))
cmat.columns = rows_name
cmat["act"] = rows_name
cmat
```

|   | pos | neg | neu | act |
|---|-----|-----|-----|-----|
| 0 | 39110 | 4768 | 506 | pos |
| 1 | 3514 | 4402 | 324 | neg |
| 2 | 2936 | 1159 | 124 | neu |

*Figure 3-4.*

As you can see, you have enough scope to improve the algorithm. In order to understand this further, let's see some errors and explore the causes and some steps to improve the model further. From the lexicon-based approach you are now transitioning to a lexicon-and-rules-based approach. As you analyze the errors you will be coming up with rules to solve the sentiment problem.

# Approach 2: Rules-Based Approach

The next method is to improve the performance of the lexicon-based approach by adding rules. Rules are based on language patterns and the structure of the data. In order to identify the rules, you first investigate the errors by comparing the predicted sentiment and the actual tags. See Listing 3-9.

*Listing 3-9.*

```
pd.options.display.max_colwidth=1000
t3.loc[t3.score_bkt!=t3.final_tags,["Summary","full_txt","final_
tags","score_bkt","pos_percent","neg_percent","pos_set","neg_set"]]
```

The following are the observations derived from the output of Listing 3-9.

# Observation 1

The sentiment in the Summary column is very clear, concise, and explicit. You can use it in your favor and weigh the hits of Summary more than full_txt. See Table 3-1.

*Table 3-1.*

| Summary | full_txt | pos_set | neg_set |
| --- | --- | --- | --- |
| My fickle cats love them | my fickle cats love them i must agree with the other 2 reviews. my cats are so-o-o-o-o ... | {like, excited, favorite, love} | {bored, refuse, picky, fickle} |
| My dogs loving it..... | my dogs loving it..... my dogs have enjoyed it for many years... | {enjoyed, loving} | {problem} |
| My daughter loves! | my daughter loves! my daughter loves these... | {loves} | {cheesy} |
| good for a dog if i have one | good for a dog if i have one i love jack link beef jerky especially the peppered flavor one. here .... | {honest, love, good, great, like} | {hard, wrong, jerky} |

# Observation 2

Booster words like "very" and "extreme" accentuate the emotion more. An example is shown in Table 3-2.

*Table 3-2. Booster Words*

| Summary | full_txt | pos_set | neg_set |
| --- | --- | --- | --- |
| Very helpful with acid reflux | very helpful with acid reflux this formula isn't the same as adding rice cereal to formula or breast milk.... | {helpful, recommend, easy} | {refused, thicker, refusing} |

You will be identifying the booster words and combining them with the positive words and seeing if there is a match. If there is a match, you will weigh it in favor of that polarity.

# Observation 3

A negation word is the negative of a given word. It can be positive or negative. "Did not like" is a negation instance. Since your lexicons contain only a single word, if you do not take care of negation words you could tag the sentence with the wrong polarity. Similar to the booster word approach, you will identify a set of negative words and cross-reference them with a negative and positive set to see if there is an occurrence of negation in the sentence. If there is an occurrence, you will subtract the score from that polarity. Let's say you have five positive hits and two negation hits. You will subtract the scores of 5 and 3. You can also give a higher weight to negation than the positive hits. See Table 3-3.

*Table 3-3.*

| Summary | full_txt | pos_set | neg_set |
|---|---|---|---|
| Not just a pepper flavor | not just a pepper flavor i was really excited to have a blend of peppercorns however since there is allspice mixed in with these it really throws the flavor off of well all of my dishes you can taste the pepper yet you also taste a earthy kind of pungent bite as well i even tried picking through the peppercorns and removing the allspice it was too late the flavor is still mixed in there <shrug> too bad | {excited, well} | {bad} |
| Not Odor Free as Advertised | not odor free as advertised when i began purchasing these.... | {appreciate, free, improves, satisfied} | {unfortunate, odor, crazy, bully, smell} |

# Observation 4

Exclamation marks are another significant token through which you can understand the user's emotion. Exclamation marks on their own do not convey a positive or a negative emotion. They only accentuate the underlying emotion. For instance "Bad!" and "Delicious!" are followed by exclamation marks but one stands for a negative experience and the other stands for positive experience. See Table 3-4.

***Table 3-4.** Exclamation Marks*

| Summary | full_txt | pos_set | neg_set |
|---------|----------|---------|---------|
| The only nut thin flavor that is dairy-free! | the only nut thin flavor that is dairy-free! my favorite flavor because it is also dairy free and even though they are not as strongly flavored they are delicious with hummus! this is my favorite gluten-free snack | {free, favorite, delicious} | { } |
| love it!!!!!!!!! | love it!!!!!!!!! since i have been diagnosed as a celiac it has become... | {love, enjoy, thank} | { } |

# Overall Score

You will compute four overall scores: positive hits on the full_txt, negative hits on the full_txt, positive hits on Summary, negative hits on Summary. You then combine them to get a final score as described Figure 3-5.

> Positive Score=number of positive hits + number of booster positive hits + number of exclamation positive hits - 2* negation (positive) hits
>
> Negative Score =number of negative hits + number of booster negative hits + number of exclamation negative hits - 2* negation (negative) hits

***Figure 3-5.** Score equations*

You apply the formula in Figure 3-5 for both full_txt and Summary and then use the formula in Figure 3-6 to get positive and negative scores.

> Score positive = Score positive for full text + 1.5* score positive for summary
>
> Score negative =Score negative for full text + 1.5* score negative for summary

***Figure 3-6.** Equations for postive and negative scores*

You now compare all four scores and get a final tag based on the flowchart in Figure 3-7.

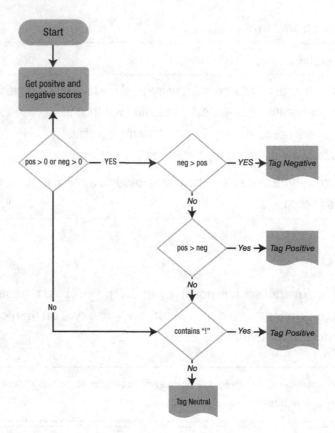

*Figure 3-7.* *Flowchart for the final tag*

Based on the flowchart, you first check if the positive or negative scores are greater than zero. If so, you further check if positive is greater or negative is greater. Accordingly, you label the final tags as positive or negative. If this is not the case, then you check for an intensity indicator, which in your case is the exclamation mark. The presence of the exclamation is a strength indicator and you know it should be either positive or negative. I have marked those cases as positive (they could be decided as negative as well depending on accuracy and the dataset at hand). The rest are marked as neutral. In cases where there are no positive or negative hits, you again check for an exclamation mark and tag it as positive or neutral.

# Implementing the Observations

## Preprocessing

You will now start implementing the observations in Python. As a first step, you want to do some preprocessing. This is especially useful for handling two-word negations. The idea is to replace bigrams like "did not" and "would not" with a single word like "didn't" and "wouldn't." So the methods you adopted for finding and replacing continue to work even when you are actually looking for two-word combinations. First, you define the bigram words that need to be replaced and then replace them with a single word, as in Listing 3-10.

***Listing 3-10.*** Replacing Two-Word Combinations

```
not_list = ["did not","could not","cannot","would not","have not"]

def repl_text(t3,col_to_repl):

    t3[col_to_repl] = t3[col_to_repl].str.replace("'","")
    t3[col_to_repl] = t3[col_to_repl].str.replace('[.,]+'," ")
    t3[col_to_repl] = t3[col_to_repl].str.replace("[\s]{2,}"," ")

    for i in not_list:
        repl = i.replace("not","").lstrip().rstrip() + "nt"
        repl = repl + " "
        t3[col_to_repl] = t3[col_to_repl].str.replace(i,repl)

    return t3
```

You now call this function for both columns, full_txt and Summary. You are going to apply all of your functions on both columns to get the four scores you saw earlier:

```
t3 = repl_text(t3,"full_txt")
t3 = repl_text(t3,"Summary")
```

## Booster and Negation Words (Observation 2 and Observation 3)

Now you define the booster and negation words. They need to be single words. If they are two words, they have to be handled in the preprocessing repl_text function. The idea behind this is to combine the booster word with a subset of positive words (pos_set1) that had a match with that sentence and see the combination. You have a counter to count both the positive and negative booster hits in this way. See Listing 3-11.

***Listing 3-11.***

```
booster_words = set(["very","extreme","extremely","huge"])

def booster_chks(com_boost,pos_set1,neg_set1,full_txt_str):
    boost_num_pos = 0
    boost_num_neg = 0
    if(len(pos_set1)>0):
        for a in list(com_boost):
            for b in list(pos_set1):
                wrd_fnd = a + " " + b
                #print (wrd_fnd)
                if(full_txt_str.find(wrd_fnd)>=0):
                    #print (wrd_fnd,full_txt_str,"pos")
                    boost_num_pos = boost_num_pos +1
    if(len(neg_set1)>0):
        for a in list(com_boost):
            for b in list(neg_set1):
                wrd_fnd = a + " " + b
                if(full_txt_str.find(wrd_fnd)>=0):
                    #print (wrd_fnd,full_txt_str,"neg")
                    boost_num_neg = boost_num_neg +1

    return boost_num_pos,boost_num_neg
```

Next, you replicate the process for negation words. See Listing 3-12.

***Listing 3-12.***

```
negation_words = set(["no","dont","didnt","cant","couldnt"])

def neg_chks(com_negation,pos_set1,neg_set1,full_txt_str):
    neg_num_pos = 0
    neg_num_neg = 0
    if(len(pos_set1)>0):
        for a in list(com_negation):
            for b in list(pos_set1):
                wrd_fnd = a + " " + b
                #print (wrd_fnd)
```

```
        if(full_txt_str.find(wrd_fnd)>=0):
            #print (wrd_fnd,full_txt_str,"pos")
            neg_num_pos = neg_num_pos +1

    if(len(neg_set1)>0):
        for a in list(com_negation):
            for b in list(neg_set1):
                wrd_fnd = a + " " + b
                if(full_txt_str.find(wrd_fnd)>=0):
                    #print (wrd_fnd,full_txt_str,"neg")
                    neg_num_neg = neg_num_neg +1

    return neg_num_pos,neg_num_neg
```

Let's do a small test to demonstrate the function. You pass the matched negation word with the sentence, matched positive set, matched negative set (in this case, it's an empty set), and the sentence. You see that you get a hit of 1 negation of a positive word ("didn't like"). See Listing 3-13.

***Listing 3-13.***

```
str_test ="it was given as a gift and the receiver didnt like it i wished
i had bought some other kind i was believing that ghirardelli would be the
best you could buy but not when it comes to peppermint bark this is not
their best effort."
neg_chks({"didnt"},{"like", "best"},{},str_test)
(1, 0)
```

## Exclamation Marks (Observation 4)

Next you come to exclamation marks. Exclamation marks increase the intensity of the underlying sentiment, either positive or negative. You want to check if there is an exclamation mark in a sentence and if the sentence also has a positive or negative hit. You check this at the sentence level, not at the review level. You will do this for both Summary and full_txt as in other cases. See Listing 3-14.

***Listing 3-14.***  Checking for Exclamation Marks

```
def excl(pos_set1,neg_set1,full_txt_str):
    excl_pos_num=0
    excl_neg_num=0

    tok_sent = sent_tokenize(full_txt_str)
    for i in tok_sent:
        if(i.find('!')>=0):
            com_set = set(i.split()) & pos_set1
            if(len(com_set)>0):
                excl_pos_num= excl_pos_num+1
            else:
                com_set1 =set(i.split()) & neg_set1
                if(len(com_set1)>0):
                    excl_neg_num= excl_neg_num+1
    return excl_pos_num,excl_neg_num
```

## Evaluation of full_txt and Summary (Observation 1)

Listing 3-15 shows the code that first gets the set of matched positive and negative words. This part is similar to what you saw earlier. Then it makes the function calls to get a count of booster words, negation words, and exclamation words for the full_txt and Summary columns. The code then sums up based on the formula discussed in Figure 3-5 and Figure 3-6 followed by the logic in Figure 3-7 for the final label for the review to get final_tag.

There is a lot of list initiation in Listing 3-15. You want to log all the values that result in the overall score for the full_txt and Summary columns. You need this step to later analyze the results and optimize the results further.

***Listing 3-15.***

```
final_tag_list = []
pos_score_list = []
neg_score_list = []
pos_set_list = []
neg_set_list = []
```

```
neg_num_pos_list = []
neg_num_neg_list = []
boost_num_pos_list = []
boost_num_neg_list = []
excl_num_pos_list =[]
excl_num_neg_list =[]

pos_score_list_sum = []
neg_score_list_sum = []
pos_set_list_sum = []
neg_set_list_sum = []
neg_num_pos_list_sum = []
neg_num_neg_list_sum = []
boost_num_pos_list_sum = []
boost_num_neg_list_sum = []
excl_num_pos_list_sum =[]
excl_num_neg_list_sum =[]
```

In Listing 3-16, each row of the t3 dataframe is evaluated based on the steps discussed so far.

***Listing 3-16.*** Evaluating the t3 Dataframe

```
for i,row in t3.iterrows():
    full_txt_str  = row["full_txt"]
    full_txt_set = set(full_txt_str.split())
    sum_txt_str = row["Summary"].lower()

    summary_txt_set = set(sum_txt_str.split())

    sent_len = len(full_txt_set)

    ####Postive and Negative sets

    pos_set1 = (full_txt_set) & (pos_set)
    neg_set1 = (full_txt_set) & (neg_set)

    com_pos_sum = (summary_txt_set) & (pos_set)
    com_neg_sum = (summary_txt_set) & (neg_set)
```

```
####Booster and Negation sets
com_boost = (full_txt_set) & (booster_words)
com_negation = (full_txt_set) & (negation_words)
com_boost_sum = (summary_txt_set) & (booster_words)
com_negation_sum = (summary_txt_set) & (negation_words)

boost_num_pos=0
boost_num_neg=0
neg_num_pos=0
neg_num_neg =0
excl_pos_num=0
excl_neg_num = 0

boost_num_pos_sum=0
boost_num_neg_sum=0
neg_num_pos_sum=0
neg_num_neg_sum =0
excl_pos_num_sum=0
excl_neg_num_sum = 0

####Get counters for booster,negation and exclamation sets
if(len(com_boost)>0):
    boost_num_pos,boost_num_neg = booster_chks(com_boost,pos_set1,neg_
    set1,full_txt_str)
    boost_num_pos_sum,boost_num_neg_sum = booster_chks(com_boost_
    sum,com_pos_sum,com_neg_sum,sum_txt_str)

if(len(com_negation)>0):
    neg_num_pos,neg_num_neg = neg_chks(com_negation,pos_set1,neg_
    set1,full_txt_str)
    neg_num_pos_sum,neg_num_neg_sum = neg_chks(com_negation_sum,com_
    pos_sum,com_neg_sum,sum_txt_str)

if((full_txt_str.find("!")>=0) & ((neg_num_pos+neg_num_neg)==0)):
    excl_pos_num,excl_neg_num = excl(pos_set1,neg_set1,full_txt_str)
    excl_pos_num_sum,excl_neg_num_sum = excl(com_pos_sum,com_neg_
    sum,sum_txt_str)
```

```
####Compute overall scores

score_pos =  len(pos_set1) + boost_num_pos - 2*neg_num_pos + 2*excl_
pos_num
score_pos_sum = len(com_pos_sum) + boost_num_pos_sum - 2*neg_num_pos_
sum + 2*excl_pos_num_sum

score_pos = score_pos + 1.5*score_pos_sum

score_neg = len(neg_set1) + 3*len(com_neg_sum) + boost_num_neg - 2*neg_
num_neg + 2*excl_neg_num
score_neg_sum = len(com_neg_sum) + boost_num_neg_sum - 2*neg_num_neg_
sum + 2*excl_neg_num_sum

score_neg = score_neg + 1.5*score_neg_sum

####Final Decision
if((score_pos>0)|(score_neg>0)):
    if((score_neg>score_pos)):
        final_tag = "neg"
    elif(score_pos>score_neg):
        final_tag = "pos"
    else:
        if(full_txt_str.find("!")>=0):
            final_tag = "pos"
        else:
            final_tag = "neu"

else:
    if(full_txt_str.find("!")>=0):
        final_tag="pos"
    else:
        final_tag="neu"

####Log intermitent values for troubleshooting

final_tag_list.append(final_tag)
pos_score_list.append(score_pos)
neg_score_list.append(score_neg)
```

```
    pos_set_list.append(pos_set1)
    neg_set_list.append(neg_set1)
    neg_num_pos_list.append(neg_num_pos)
    neg_num_neg_list.append(neg_num_neg)
    boost_num_pos_list.append(boost_num_pos)
    boost_num_neg_list.append(boost_num_neg)
    excl_num_pos_list.append(excl_pos_num)
    excl_num_neg_list.append(excl_neg_num)

    pos_set_list_sum.append(com_pos_sum)
    neg_set_list_sum.append(com_neg_sum)
    neg_num_pos_list_sum.append(neg_num_pos_sum)
    neg_num_neg_list_sum.append(neg_num_neg_sum)
    boost_num_pos_list_sum.append(boost_num_pos_sum)
    boost_num_neg_list_sum.append(boost_num_neg_sum)
    excl_num_pos_list_sum.append(excl_pos_num_sum)
    excl_num_neg_list_sum.append(excl_neg_num_sum)
```

Listing 3-17 sets all the values to your dataframe for understanding accuracies and identifying any opportunity for improvement.

***Listing 3-17.***

```
t3["final_tags"] = final_tag_list
t3["pos_score"] = pos_score_list
t3["neg_score"] = neg_score_list

t3["pos_set"] = pos_set_list
t3["neg_set"] = neg_set_list

t3["neg_num_pos_count"] = neg_num_pos_list
t3["neg_num_neg_count"] = neg_num_neg_list

t3["boost_num_pos_count"] = boost_num_pos_list
t3["boost_num_neg_count"] = boost_num_neg_list

t3["pos_set_sum"] = pos_set_list_sum
t3["neg_set_sum"] = neg_set_list_sum
```

```
t3["neg_num_pos_count_sum"] = neg_num_pos_list_sum
t3["neg_num_neg_count_sum"] = neg_num_neg_list_sum

t3["boost_num_pos_count_sum"] = boost_num_pos_list_sum
t3["boost_num_neg_count_sum"] = boost_num_neg_list_sum

t3["excl_num_pos_count"] = excl_num_pos_list
t3["excl_num_neg_count"] = excl_num_neg_list

t3["excl_num_pos_count_sum"] = excl_num_pos_list_sum
t3["excl_num_neg_count_sum"] = excl_num_neg_list_sum
```

In Listing 3-18, you compute accuracies and the confusion matrix and F1 score.

***Listing 3-18.*** Accuracies, Confusion Matrix, and F1 Score

```
from sklearn.metrics import accuracy_score
print (accuracy_score(t3["score_bkt"],t3["final_tags"]))
0.8003448093872596

from sklearn.metrics import f1_score
f1_score(t3["score_bkt"],t3["final_tags"], average='macro')
0.502500231042986
```

You can see an improvement in the numbers in Listing 3-18. Let's check on the confusion matrix in Listing 3-19 and Figure 3-8.

***Listing 3-19.*** The Confusion Matrix

```
rows_name = t3["score_bkt"].unique()
from sklearn.metrics import confusion_matrix
cmat = pd.DataFrame(confusion_matrix(t3["score_bkt"],t3["final_tags"],
labels=rows_name, sample_weight=None))
cmat.columns = rows_name
cmat["act"] = rows_name
cmat
```

|   | pos | neg | neu | act |
|---|-----|-----|-----|-----|
| 0 | 41060 | 2626 | 746 | pos |
| 1 | 3471 | 4169 | 487 | neg |
| 2 | 2992 | 1027 | 265 | neu |

***Figure 3-8.*** *The confusion matrix*

As you can see, the negative and positive classifications are much better as compared to neutral. This is getting your overall F1 score (Macro) down. However, in neutral, you are doing pretty badly. You can also recheck your scores by removing neutral from your population. See Listing 3-20.

***Listing 3-20.*** Removing Neutral from Your Population

```
t4 = t3.loc[(t3.score_bkt!="neu") & (t3.final_tags!="neu")].reset_index()
print (accuracy_score(t4["score_bkt"],t4["final_tags"]))
print (f1_score(t4["score_bkt"],t4["final_tags"],average='macro'))
0.8812103027705257
0.75425508403348
```

## Optimizing the Code

Clearly your strategy is working for positive and negative classification. So, in the next iteration you should focus and optimize for neutral sentiment. The following are some strategies for the next set of optimizations (for both neutral and otherwise). You can take this problem forward and solve it with any strategies you may think up and use the current scores as the benchmark.

- **Capitalization**: Words that are capitalized indicate intensity similar to exclamation. Sometimes capitalized words are sentiment words in themselves or they can be adjacent to a sentiment word, such as "EXTREMELY happy" or "EXTREMELY angry with the product." You need to identify the capitalized words accordingly and tweak the score accordingly.

- **Lexicons**: You can try with different lexicons. There are also lexicons that provide weights to different sentiment words. The paper titled "A lexicon based method to search for extreme opinions" at `https://journals.plos.org/plosone/article?id=10.1371/journal.pone.0197816` describes a method to generate lexicons from existing online ratings. The idea is to use the words present in a given review with a rating as source of a lexicon and then normalize these ratings across all reviews to arrive at a lexicon with a weight scale to it. This lexicon can then be used for any classification purposes.

- **Use semantic orientation for determining polarity:** Another approach is to use the Web or any other external source to determine the polarity of the word at hand. For example, you can see how many page hits the word "delight" has together with the word "good." You can also see the number of page hits the word "delight" has together with the word "bad." You can compare the results to know whether the word "delight" should have a positive or a negative polarity. The following method is from the paper titled "Thumbs up or thumbs down?: semantic orientation applied to unsupervised classification of reviews" at `https://dl.acm.org/doi/10.3115/1073083.1073153`. The similarity between the known positive word (e.g. "good") and the given lexicon is measured based on point-wise mutual information as well as the Jaccard Index. The formulas are explained in Figure 3-9 and Figure 3-10.

$$PMI = log\ (P(word1) * P(word2)/P(word1, word2))$$

*Figure 3-9.*

$$Jaccard = P(word1\ \&\ Word2)/(P(Word1) + P(Word2) - P(Word1\ \&\ Word2))$$

*Figure 3-10.*

Here P(Word1) or P(Word2) refers to the number of hits the word has in the web search engine. P(Word1 & Word2) is the number of hits the word has for word1 and word2 in the same query. Listings 3-21 and 3-22 demonstrated the approach for few words. You get semantic orientation using web search results between a given word and a phrase of interest. This can be scaled for all words in a lexicon. This can also be done for the adjectives in sentences in the corpus. For the following exercise, you need a Bing API key to get the number of hits in the Bing search engine. The steps to create a Bing API key are as follows:

1. Go to `https://azure.microsoft.com/en-us/try/cognitive-services/`.

2. Click the Search APIs tab in the screen in Figure 3-11.

***Figure 3-11.***  *Searching for APIs*

3. Click the Get API Key (Bing Search APIs V7) button shown in Figure 3-12.

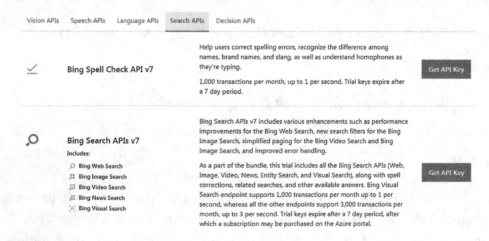

***Figure 3-12.***  *Getting the API key*

4.  You will be asked to log in and you will see a page similar to Figure 3-13. Scroll down to get Key 1.

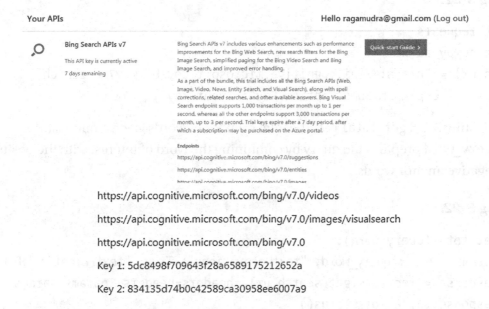

Your APIs                                                      Hello ragamudra@gmail.com (Log out)

Bing Search APIs v7                    Bing Search APIs v7 includes various enhancements such as performance
                                       improvements for the Bing Web Search, new search filters for the Bing
This API key is currently active       Image Search, simplified paging for the Bing Video Search and Bing
                                       Image Search, and improved error handling.                          Quick-start Guide >
7 days remaining
                                       As a part of the bundle, this trial includes all the Bing Search APIs (Web,
                                       Image, Video, News, Entity Search, and Visual Search), along with spell
                                       corrections, related searches, and other available answers. Bing Visual
                                       Search endpoint supports 1,000 transactions per month up to 1 per
                                       second, whereas all the other endpoints support 3,000 transactions per
                                       month, up to 3 per second. Trial keys expire after a 7 day period, after
                                       which a subscription may be purchased on the Azure portal.

                                       Endpoints
                                       https://api.cognitive.microsoft.com/bing/v7.0/suggestions
                                       https://api.cognitive.microsoft.com/bing/v7.0/entities
                                       https://api.cognitive.microsoft.com/bing/v7.0/images

https://api.cognitive.microsoft.com/bing/v7.0/videos

https://api.cognitive.microsoft.com/bing/v7.0/images/visualsearch

https://api.cognitive.microsoft.com/bing/v7.0

Key 1: 5dc8498f709643f28a6589175212652a

Key 2: 834135d74b0c42589ca30958ee6007a9

***Figure 3-13.***  *Getting Key 1*

Once you have the Bing API, follow these steps:

1.  Get the number of hits of anchor words. For positive, you consider the word "good" and for negative, you consider the word "bad."

2.  For each word in the list of words of interest, get the total number of hits.

3.  For each word, get the total hits of the word and each of the anchor words.

4.  Compute the PMI and the Jaccard. When the number of hits is below a threshold, you invalidate the score. This is because a good part of the hits could be mere chance. Since these are web searches, you will keep the threshold at a few millions. This is based on the paper titled "Measuring semantic similarity between words using web search engines" at `http://www2007.org/papers/paper632.pdf`.

5.  Compare the scores and tag the polarity.

Let's get started. See Listing 3-21.

***Listing 3-21.***

```
import requests
import numpy as np
search_url = "https://api.cognitive.microsoft.com/bing/v7.0/search"
headers = {"Ocp-Apim-Subscription-Key": "your-key"}
```

In Listing 3-22, get_total gets the total number of words given a query and get_query_word prepares the query by combining the word of interest with the positive and negative anchor words.

***Listing 3-22.***

```
def get_total(query_word):
    params = {"q": query_word, "textDecorations": True, "textFormat": "HTML"}
    response = requests.get(search_url, headers=headers, params=params)
    response.raise_for_status()
    search_results = response.json()

    return search_results['webPages']['totalEstimatedMatches']
def get_query_word(str1,gword,bword):
    str_base = gword
    str_base1 =bword
    query_word_pos = str1 + "+" + str_base
    query_word_neg = str1 + "+" + str_base1

    return query_word_pos,query_word_neg
```

In Listing 3-23, the get_pmi and get_jaccard functions gets the pointwise mutual information score and the Jaccard score for all input parameters based on the formula described Figure 3-10. If the number of hits is insufficient, they're tagged as "na."

***Listing 3-23.***

```
def get_pmi(hits_good,hits_bad,hits_total,sr_results_pos_int,sr_results_
neg_int,str1_tot):

    pos_score = "na"
    neg_score = "na"

    if(sr_results_pos_int>=1000000):
        pos_score = np.log((hits_total*sr_results_pos_int)/(hits_good*
        str1_tot))
    if(sr_results_neg_int>=1000000):
        neg_score = np.log((hits_total*sr_results_neg_int)/(str1_tot*
        hits_bad))

    return pos_score,neg_score
def get_jaccard(hits_good,hits_bad,sr_results_pos_int,sr_results_neg_int,
str1_tot):

    pos_score = "na"
    neg_score = "na"

    if(sr_results_pos_int>=1000000):
        pos_score = sr_results_pos_int/(((str1_tot+hits_good)-sr_results_
        pos_int))
    if(sr_results_neg_int>=1000000):
        neg_score = sr_results_neg_int/(((str1_tot+hits_bad)-sr_results_
        neg_int))

    return pos_score,neg_score
```

You now define the anchor words and words of interest for which you want to understand polarity (list1). In the case of anchor words, you want to understand the total number of hits when either of these words are present (you define this space as the universe). This is needed to compute PMI and Jaccard Scores. The other thing to note is that since you are interested in words of sentiment that are typically expressed in reviews, you add the word "reviews" to the scope your search. This reigns in the context to sentiment. These are optimizations that I have applied. Please feel to modify as per the problem at hand. See Listing 3-24.

*Listing 3-24.*

```
gword = "good"
bword = "bad"

hits_good =get_total(gword)
hits_bad =get_total(bword)
hits_total = get_total(gword + " OR " + bword)

list1= ["delight","pathetic","average","awesome","tiresome","angry","furious"]
```

Now you iterate through this list and determine the polarity for each of these words. The final outputs can be positive, negative, or indeterminate. See Listing 3-25.

*Listing 3-25.*  Determining Polarity

```
for i in list1:
    str1 = i + "+reviews"
    str1_tot = get_total(str1)

    query_word_pos,query_word_neg =get_query_word(str1,gword,bword)
    sr_results_pos_int = get_total(query_word_pos)
    sr_results_neg_int = get_total(query_word_neg)
    pmi_score =  get_pmi(hits_good,hits_bad,hits_total,sr_results_pos_
    int,sr_results_neg_int,str1_tot)
    jc_score = get_jaccard(hits_good,hits_bad,sr_results_pos_int,sr_
    results_neg_int,str1_tot)

    if(pmi_score[0]=="na" or pmi_score[1]=="na"):
        print (i,"pmi indeterminate")

    elif(pmi_score[0]>pmi_score[1]):
        print (i,"pmi pos")

    elif(pmi_score[0]<pmi_score[1]):
        print (i,"pmi neg")

    else:
        print (i,"pmi neutral")
```

```
if(jc_score[0]=="na") or (jc_score[1]=="na"):
    print (i,"jc indeterminate")

elif(jc_score[0]>jc_score[1]):
    print (i,"jc pos")

elif(jc_score[0]<pmi_score[1]):
    print (i,"jc neg")
else:
    print (i,"jc neutral")
```

The output is shown in Listing 3-26. Some of the output, for instance like "average," is coming out as positive since you did not define a neutral state. You could add a condition that if the difference between positive and negative is within a threshold, that would amount to the word being neutral.

***Listing 3-26.*** The Output

```
delight pmi pos
delight jc pos
pathetic pmi neg
pathetic jc neg
average pmi pos
average jc pos
awesome pmi indeterminate
awesome jc indeterminate
tiresome pmi indeterminate
tiresome jc indeterminate
angry pmi neg
angry jc neg
furious pmi neg
furious jc ncg
```

You could also use another corpus like a wikipedia or a news corpus. You could try different combinations of anchor words and also some other scoping words (in your case, the word "reviews") to achieve better results.

# Sentiment Analysis Libraries

There are also standard libraries you can use to do sentiment analysis. They offer out-of-the-box sentiment analysis and hence must be used appropriately. If it fits the problem at hand, then you have a quick and robust solution. You could also use it along with the approach/algorithm that they use or as a standalone solution. Vader (Valence Aware Dictionary and Sentiment Reasoner) is a Python library that does sentiment analysis and provides the polarity of a word with a weight. The original research paper is titled "Vader: A parsimonious rule-based model for sentiment analysis of social media text" and is at `https://www.aaai.org/ocs/index.php/ICWSM/ICWSM14/paper/viewPaper/8109`). Fundamentally Vader took set of words from well-established lexicons as well as from a social media corpus. Each of the words was provided a polarity score by a set of people following a wisdom-of-crowd approach to arrive at final lexicons and their weights. Listing 3-27 shows how to install the Vader library using `pip install`.

***Listing 3-27.*** Installing the Vader Library

```
!pip install vaderSentiment
```

Listing 3-28 provides a quick way to use the Vader library.

***Listing 3-28.*** Using the Vader Library

```
from vaderSentiment.vaderSentiment import SentimentIntensityAnalyzer
analyser = SentimentIntensityAnalyzer()
score = analyser.polarity_scores("I am good")
score
{'neg': 0.0, 'neu': 0.256, 'pos': 0.744, 'compound': 0.4404}
```

The compound score in this example is the final weighted scores of the polarity of that sentence. Positive and negative scores are the percentage of lexicons that fall in that polarity. They will sum up to 1.

# Approach 3: Machine-Learning Based Approach (Neural Network)

Since you have the final rating provided by the user, you can try fitting a supervised model. One way to do a supervised model in your case is to use all of the text from the Summary and Text columns as features, score_bkt as the target variable, and fit a machine learning model. The one issue with this approach is that the model could get biased by the product names or a product category that had a higher or a lower rating. The other way is to use the lexical features you created earlier as features in the model. Table 3-5 shows the intermediary variables you created in the lexical-based approach. You will add few more variables to this and build your model.

***Table 3-5.*** *Intermediary Variables*

| Lexical features | Variable meaning |
| --- | --- |
| sent_len | Length of the text |
| pos_score | Positive score by the lexical method |
| neg_score | Negative score by the lexical method |
| neg_num_pos_count | Number of negative words near positive words |
| neg_num_neg_count | Number of negative words near positive words |
| boost_num_pos_count | Number of booster words near positive words |
| boost_num_neg_count | Number of booster words near negative words |
| neg_num_neg_count_sum | Number of negative words near positive words in summary |
| boost_num_pos_count_sum | Number of negative words near positive words in summary |
| boost_num_neg_count_sum | Number of booster words near positive words in summary |
| excl_num_pos_count | Number of exclamation marks near positive words |
| excl_num_neg_count | Number of exclamation marks near negative words |
| excl_num_pos_count_sum | Number of exclamation marks near positive words in summary |
| excl_num_neg_count_sum | Number of exclamation marks near negative words in summary |

# Corpus Features

With the text and summary of the corpus, you can create bag-of-words features using the TF-IDF vectorizer. Since you want to create a generic sentiment analysis module, you select only words that are relevant to emotions or sentiments words from the corpus. This is an important step since using all of the words (in your case, pet food names, candy names, etc.) could bias the model to learn patterns that associate product and their associated sentiment. You will create two sets of word features: containing only adjectives and containing only stop words. Please note in your classification example in Chapter 2 you eliminated stop words. Here you will use stop words as the only features set.

You will then combine these two features sets and with `score_bkt` as the dependent variable you will train a neural network, adjusting for class weights to get the best possible model that provides a good classification across all levels. You are using NLTK version 3.4.3 here, as mentioned in Table 3-7. See Listing 3-29 and Listing 3-30.

***Listing 3-29.***

```
import pandas as pd
from sklearn.linear_model import LogisticRegression
import nltk
import warnings
import stop_words
warnings.filterwarnings('ignore')
```

***Listing 3-30.***

```
t1 = pd.read_csv("lexicon_sent_processed.csv")
tgt = t1.loc[:,"score_bkt"]
```

The following code shows functions to get text features. The function `cnv_str` converts positive, negative set into strings so that they can be vectorized and used in the model. The function `filter_pos` filters out only adjectives from the Text and Summary sentences. The function `get_stop_words` keeps only stop_words from the text corpus. See Listings 3-31 through 3-33.

*Listing 3-31.*

```
def cnv_str(x):
    x1 = list(eval(x))
    x2 = ' '.join(x1)
    return x2
```

*Listing 3-32.*

```
def filter_pos(fltr,sent_list):
    str1 = ""
    for i in sent_list:
        if(i[1]=="JJ"):
            str1 = str1 + i[0].lower() + " "
    return str1
```

*Listing 3-33.*

```
def get_stop_words(sent):
    list1 = set(sent.split())
    st_comm = list(list1 & st_set)
    st_comm = ' '.join(st_comm)
    return st_comm
```

Listing 3-34 applies these functions to the positive/negative set and text fields.

*Listing 3-34.*

```
t1["pos_set1"] = t1["pos_set"].apply(cnv_str)
t1["neg_set1"] = t1["neg_set"].apply(cnv_str)
t1["pos_neg_comb"] = t1["pos_set1"] + " " + t1["neg_set1"]

get_pos_tags = nltk.pos_tag_sents(t1["Text"].str.split())

str_sel_list = []
for i in get_pos_tags:
    str_sel = filter_pos("JJ",i)
    str_sel_list.append(str_sel)
```

```
t1["pos_neg_comb_adj"] = t1["pos_neg_comb"] + str_sel_list
```

```
st1 = stop_words.get_stop_words('en')
st_set = set(st1)
onl_stop_words = t1["full_txt"].apply(get_stop_words)
```

```
t1["pos_neg_comb_adj_st"] = t1["pos_neg_comb_adj"] + onl_stop_words
```

Listing 3-35 shows a stratified split for making the train and test datasets in order to get sample accuracies.

### Listing 3-35.

```
from sklearn.model_selection import StratifiedShuffleSplit
sss = StratifiedShuffleSplit(test_size=0.8,random_state=42,n_splits=1)

for train_index, test_index in sss.split(t1, tgt):
    x_train, x_test = t1[t1.index.isin(train_index)], t1[t1.index.
    isin(test_index)]
    y_train, y_test = t1.loc[t1.index.isin(train_index),"score_bkt"],
    t1.loc[t1.index.isin(test_index),"score_bkt"]
```

You keep only lexicon features in Listing 3-36. The text features will be processed in another step and concatenated with the numeric arrays: x_train1 and x_test1. See Figure 3-14.

### Listing 3-36.

```
inde_vars = ["sent_len", "pos_score", "neg_score", "neg_num_pos_count",
"neg_num_neg_count", 'boost_num_pos_count', 'boost_num_neg_count',
'neg_num_pos_count_sum', 'neg_num_neg_count_sum', 'boost_num_pos_count_sum',
'boost_num_neg_count_sum', 'excl_num_pos_count', 'excl_num_neg_count',
'excl_num_pos_count_sum', 'excl_num_neg_count_sum']
x_train1 = x_train[inde_vars]
x_test1 = x_test[inde_vars]
x_train1.head()
```

| | sent_len | pos_score | neg_score | neg_num_pos_count | neg_num_neg_count | boost_num_pos_count | b |
|---|---|---|---|---|---|---|---|
| 5 | 42.0 | 3.0 | 0.0 | 0 | 0 | 0 | |
| 7 | 38.0 | 1.0 | 5.5 | 0 | 0 | 0 | |
| 40 | 97.0 | 8.0 | 0.0 | 0 | 1 | 0 | |
| 47 | 45.0 | 8.5 | 0.0 | 0 | 0 | 1 | |
| 48 | 88.0 | 2.0 | 3.0 | 0 | 0 | 0 | |

*Figure 3-14.*

Listing 3-36 showing the numeric features in the dataset. You now vectorize and create matrices out of columns containing adjectives, stop words, and positive and negative matched words. See Listing 3-37.

*Listing 3-37.*

```
from sklearn.feature_extraction.text import TfidfVectorizer

tfidf_vectorizer = TfidfVectorizer(min_df=0.001,analyzer=u'word',
ngram_range=(1,1))
tfidf_matrix_tr = tfidf_vectorizer.fit_transform(x_train["pos_neg_comb_adj_st"])

tfidf_matrix_te = tfidf_vectorizer.transform(x_test["pos_neg_comb_adj_st"])

x_train2= tfidf_matrix_tr.todense()
x_test2 = tfidf_matrix_te.todense()
```

x_train2 and x_test2 are the matrices that processed a set of text features. This will then be concatenated with numeric matrices x_train1 and x_test2. See Listing 3-38.

*Listing 3-38.*

```
import numpy as np
x_train3 = np.concatenate([x_train1,x_train2],axis=1)
x_test3 = np.concatenate([x_test1,x_test2],axis=1)
```

Since you will be building a neural network classifier with score_bkt as the target variable, it is important to standardize the features. You are going to standardize using a min-max scaler. Listing 3-39 shows the formula of a min-max scaler using the sklearn function sklean.

```
X_std = (X - X.min(axis=0)) / (X.max(axis=0) - X.min(axis=0))
X_scaled = X_std * (max - min) + min
```

***Listing 3-39.*** The Min-Max Scaler

```
from sklearn import preprocessing
min_max_scaler = preprocessing.MinMaxScaler()
x_train_scaled = min_max_scaler.fit_transform(x_train3)
x_test_scaled = min_max_scaler.transform(x_test3)
```

You reduce the number of features using the feature selection algorithm, as you did in the last chapter. See Listing 3-40.

***Listing 3-40.***

```
from sklearn.feature_selection import SelectPercentile, f_classif
selector = SelectPercentile(f_classif, percentile=40)
selector.fit(x_train3,y_train)
x_train4 = selector.fit_transform(x_train_scaled,y_train)
x_test4 = selector.transform(x_test_scaled)
```

You also need to convert the categorical target variable into a one hot encoding form. One hot encoding is a form where each level is represented by the absence of the other levels and the presence of that level, so positive can be represented as 100, negative as 010, and neutral as 001.

***Listing 3-41.*** One Hot Encoding

```
from sklearn.preprocessing import LabelEncoder
from keras.utils import np_utils
le = LabelEncoder()
y_train1 = le.fit_transform(y_train)
y_train2 = np_utils.to_categorical(y_train1)
y_test1 = le.transform(y_test)
print (y_train2.shape)
(11368, 3)
```

# Building the Neural Network

Now that you have the independent variables and the dependent variables in the right shape, you will start your process of building a neural network. Your neural network will be a shallow network with four layers and the number of nodes in the layer will reduce as the network proceeds. You will have 500, 200, 100, 50 layers. The last layer will have three inputs with softmax activation since you are solving a multi-class problem with three classes. Please feel free to tweak the network for better results. Since you have an unbalanced class problem, you will also provide a class weight to a neural network in favor of the lower classes. See Listing 3-42 through Listing 3-45.

***Listing 3-42.***

```
import tensorflow as tf
import keras
from keras.models import Sequential
from keras.layers import Dense
from keras.layers import Input, Dense, Dropout
from keras.models import Model
from keras.utils import to_categorical
from keras.optimizers import Adam
```

***Listing 3-43.***

```
def get_nn_mod(list_layers,dp):
    model = Sequential()
    model.add(Dense(list_layers[0], input_dim=x_train4.shape[1],
    activation='tanh', kernel_initializer='lecun_uniform'))
    model.add(Dropout(dp))

    for i in list_layers[1:]:
        model.add(Dense(i, input_dim=x_train4.shape[1], activation='tanh'))
        model.add(Dropout(dp))
```

```
    model.add(Dense(3, activation='softmax'))
    opt = Adam(0.0001)
# Compile model
    model.compile(optimizer=opt, loss='categorical_crossentropy',
                  metrics=['accuracy'])
    return model
```

***Listing 3-44.***

```
list_layers = [500,200,100,50]
class_weight = {0:0.2,1:0.6,2:0.2}

model = get_nn_mod(list_layers,0.1)
```

***Listing 3-45.***

```
model.fit(x_train4,y_train2, batch_size=100, epochs=20,class_weight=class_
          weight, verbose=2,validation_split=0.2)
```

Once the model is fit, you check for the accuracy and F1 scores. You can see that the accuracy has stayed the same as the lexicon approach, but the F1 scores have improved considerably. The same can be seen with confusion matrix. See Listing 3-46, Listing 3-47, and Figure 3-15.

***Listing 3-46.***

```
pred=model.predict_classes(x_test4)
from sklearn.metrics import accuracy_score
from sklearn.metrics import f1_score
ac1 = accuracy_score(y_test1, pred)
print (ac1, f1_score(y_test1, pred, average='macro'))
0.8045519516217702 0.5858385721186434
```

***Listing 3-47.***

```
from sklearn.metrics import confusion_matrix
rows_name = t1["score_bkt"].unique()
pred_inv = le.inverse_transform(pred)
```

```
cmat = pd.DataFrame(confusion_matrix(y_test, pred_inv, labels=rows_name,
sample_weight=None))
cmat.columns = rows_name
cmat["act"] = rows_name
cmat
```

|   | pos | neg | neu | act |
|---|-----|-----|-----|-----|
| 0 | 31959 | 1033 | 2554 | pos |
| 1 | 1614 | 3378 | 1510 | neg |
| 2 | 1563 | 766 | 1098 | neu |

*Figure 3-15.*

# Things to Improve

To improve this model, you can add other text features like adverbs, conjunctions, capitalized words, and so on. The hyperparameters and class weights could also be tweaked to get a better F1 score without losing accuracy.

# Attribute Extraction

When you analyze reviews for sentiment, you will find that customers have mixed opinions about the product. For example, "The phone has a highly effective camera but the battery is pathetic." This review is very positive about the camera and extremely negative about the battery. You want to first extract attributes and then identify the polarity associated with each of the attributes. You will see a small example on attribute extraction. PromptCloud (`www.promptcloud.com/`) extracted 400K Amazon reviews of unlocked phones. Details of the dataset can be found at `https://data.world/promptcloud/amazon-mobile-phone-reviews`. This dataset has phone reviews of various handsets and brands. Your objective in this exercise is to extract the attributes from each of the reviews and then correlate the same with positive and negative ratings. You want to understand top-rated attributes for each of the brands as compared to other brands. As the first step, you will extract the attributes and then correlate with the ratings. See Listing 3-48 and Figure 3-16.

*Listing 3-48.*

```
import pandas as pd
import nltk
pd.options.display.max_colwidth=1000
t1 = pd.read_csv("Amazon_Unlocked_Mobile.csv")
t1.shape
(413840, 6)
t1.head()
```

| | Product Name | Brand Name | Price | Rating | Reviews | Review Votes |
|---|---|---|---|---|---|---|
| 0 | "CLEAR CLEAN ESN" Sprint EPIC 4G Galaxy SPH-D700*FRONT CAMERA*ANDROID*SLIDER*QWERTY KEYBOARD*TOUCH SCREEN | Samsung | 199.99 | 5 | I feel so LUCKY to have found this used (phone to us & not used hard at all), phone on line from someone who upgraded and sold this one. My Son liked his old one that finally fell apart after 2.5+ years and didn't want an upgrade!! Thank you Seller, we really appreciate it & your honesty re: said used phone.I recommend this seller very highly & would but from them again!! | 1.0 |
| 1 | "CLEAR CLEAN ESN" Sprint EPIC 4G Galaxy SPH-D700*FRONT CAMERA*ANDROID*SLIDER*QWERTY KEYBOARD*TOUCH SCREEN | Samsung | 199.99 | 4 | nice phone, nice up grade from my pantach revue. Very clean set up and easy set up. never had an android phone but they are fantastic to say the least. perfect size for surfing and social media. great phone samsung | 0.0 |
| 2 | "CLEAR CLEAN ESN" Sprint EPIC 4G Galaxy SPH-D700*FRONT CAMERA*ANDROID*SLIDER*QWERTY KEYBOARD*TOUCH SCREEN | Samsung | 199.99 | 5 | Very pleased | 0.0 |
| 3 | "CLEAR CLEAN ESN" Sprint EPIC 4G Galaxy SPH-D700*FRONT CAMERA*ANDROID*SLIDER*QWERTY KEYBOARD*TOUCH SCREEN | Samsung | 199.99 | 4 | It works good but it goes slow sometimes but its a very good phone I love it | 0.0 |
| 4 | "CLEAR CLEAN ESN" Sprint EPIC 4G Galaxy SPH-D700*FRONT CAMERA*ANDROID*SLIDER*QWERTY KEYBOARD*TOUCH SCREEN | Samsung | 199.99 | 4 | Great phone to replace my lost phone. The only thing is the volume up button does not work, but I can still go into settings to adjust. Other than that, it does the job until I am eligible to upgrade my phone again.Thaanks! | 0.0 |

*Figure 3-16.*

You want reviews that are longer, so you'll keep a minimum cutoff of at least 10 characters. Since we are doing sample work and to quickly iterate, I kept only 10% of the data. Now see Listing 3-49.

*Listing 3-49.*

```
t2 = t1[t1.Reviews.str.len()>=10]
t3 = t2.sample(frac=0.01)
```

You will follow the steps in Figure 3-17 for attributes extraction.

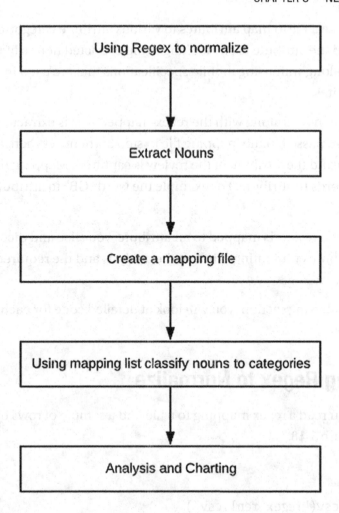

*Figure 3-17.*

1. You will use regular expressions to normalize characters or words like inches (""), dollars ($), GB, etc. I created a regular expression mapping file for the same where each of the regexes has a word that it needs to be replaced with. For example, the dollar symbol ($) can be replaced with the word "dollars." You will also extract and collect the replaced words in a separate column and this will be used in step 3.

2. You extract noun phrases here using NLTK. Attributes are typically nouns or noun forms (NN or NNP or NNS). You pass on the extracted nouns to the next step. All nouns are not attributes and hence this step acts like a filter.

3. You create a file to map attributes to various attribute categories. I created the attribute list by looking at top extracted nouns in the corpus along with using mobile specifications that are used in review sites.

4. The list of nouns along with the regex-mapped words extracted in step 1 are passed to the mapping file in step 3 whenever there is a match and the attribute in the review is captured. Mapping file maps words to attribute. For example the word "GB" to attribute "storage".

5. Once each review is mapped to an attribute, you slice and dice the attributes with ratings and get the analysis and the required outputs.

6. In the following section, you will look at detailed code for each of the steps.

## Step 1: Using Regex to Normalize

In Listing 3-50, you read a regex mapping to a file and a sample of rows of the file is provided. See Figure 3-18.

*Listing 3-50.*

```
rrpl = pd.read_csv("regex_repl.csv")
rrpl.head()
```

| | repl | word |
|---|---|---|
| 0 | [0-9.]+" | inches |
| 1 | [0-9.]+ " | inches |
| 2 | [0-9.]+ inch | inches |
| 3 | inch | inches |
| 4 | [0-9]+mm | millimeters |

*Figure 3-18.*

You follow this with a function to replace the regular expressions extracted in the sentences with corresponding words. The variable words_Coll is the set of extracted and mapped words. In case the replaced words in Reviews1 are not recognized as noun forms by the "part-of-speech" tagger we still make sure they are available for our final mapping process  by using the extracted words in words_Coll in step 3. See Listing 3-51.

***Listing 3-51.***

```
def repl_text(t3,to_repl,word_to_repl):

    t3["Reviews1"] =t3["Reviews1"].str.replace(to_repl,word_to_repl)
    ind_list = t3[t3["Reviews"].str.contains(to_repl)].index
    t3.loc[ind_list,"words_Coll"] = t3.loc[ind_list,"words_Coll"] + " " +
    word_to_repl

    return t3
```

In Listing 3-52, you run rows iteratively and replace the patterns and keep the replaced words in a separate column.

***Listing 3-52.***

```
t3["Reviews1"]  = t3["Reviews"].str.lower()
t3["words_Coll"] = ""

for index,row  in rrpl.iterrows():
    repl = row["repl"]
    word =  row["word"]
    t3 = repl_text(t3,repl,word)
```

At the end of this step you have two columns: Reviews1 (text with replaced patterns) and words_Coll (words extracted from the patterns).

# Step 2: Extracting Noun Forms

You will now extract noun forms in the following function using filter_pos. Here fltr is a variable you pass to the function. It can be reused if you want to try with other part-of-speech tags. See Listings 3-53 and 3-54.

***Listing 3-53.***

```
def filter_pos(fltr,sent_list):
    str1 = ""
    for i in sent_list:
        if(i[1].find(fltr)>=0):
            str1 = str1 + i[0].lower() + " "
    return str1
```

***Listing 3-54.***

```
get_pos_tags = nltk.pos_tag_sents(t3["Reviews1"].str.split())
str_sel_list = []
for i in get_pos_tags:
    str_sel = filter_pos("NN",i)
    str_sel_list.append(str_sel)
```

The variable `str_sel_list`, which contains noun form attributes, is combined with `words_coll`. This is to make sure all the replaced words are part of your analysis even if they get missed by the part-of-speech tagger. See Listing 3-55.

***Listing 3-55.***

```
t3["pos_tags"] = str_sel_list
t3["all_attrs_ext"] = t3["pos_tags"] + t3["words_Coll"]
```

The column `all_attrs_ext` will be used in mapping attributes to categories.

# Step 3: Creating a Mapping File

You need to create a mapping file. You could scrape mobile review websites and then start generating the mapping file. As an example, see the specifications from `www.gsmarena.com/xiaomi_redmi_k30-review-2055.php`.

- **Body:** Metal frame, Gorilla Glass 5 front and back, 208g.

- **Display:** 6.67" IPS LCD, 120Hz, HDR10, 1080 x 2400px resolution, 20:9 aspect ratio, 395ppi.

- **Rear camera:** Primary: 64MP, f/1.89 aperture, 1/1.7" sensor size, 0.8μm pixel size, PDAF. Ultra wide: 8MP, f/2.2, 1/4", 1.12μm pixels. Macro camera: 2MP, f/2.4, 1/5", 1.75μm. Depth sensor: 2MP; 2160p@30fps, 1080p@30/60/120fps, 720p@960fps video recording.

- **Front camera:** Primary: 20MP, f/2.2 aperture, 0.9μm pixels. Depth sensor: 2MP; 1080p/30fps video recording.

- **OS:** Android 10; MIUI 11.

- **Chipset:** Snapdragon 730G (8nm): Octa-core (2x2.2 GHz Kryo 470 Gold & 6x1.8 GHz Kryo 470 Silver), Adreno 618 GPU.

- **Memory:** 6/8GB of RAM; 64/128/256GB UFS 2.1 storage; shared microSD slot.

- **Battery:** 4,500mAh; 27W fast charging.

- **Connectivity:** Dual-SIM; LTE-A, 4-Band carrier aggregation, LTE Cat-12/ Cat-13; USB-C; Wi-Fi a/b/g/n/ac; dual-band GPS; Bluetooth 5.0; FM radio; NFC; IR blaster.

- **Misc:** Side-mounted fingerprint reader; 3.5mm audio jack.

For example, the rear and front could be mapped to Camera. LTE, WIFI, and so on could be mapped to Network. By looking at various review sites, I normalized the attribute categories to the following categories:

1. Communication (Comm)
2. Dimension
3. Weight
4. Build
5. Sound
6. Sim
7. Display
8. Operating system
9. Performance
10. Price

11. Battery

12. Radio

13. Camera

14. Keyboard

15. Apps

16. Warranty

17. Guarantee

18. Call quality

19. Storage

attr_cat.csv is the mapping file. A quick look at the mapping file is shown in Table 3-6.

**Table 3-6.**  *A Quick Look at the Mapping File*

| words | cat |
|---|---|
| gsm | comm |
| hspa | comm |
| lte | comm |
| 2g | comm |
| 3g | comm |
| 4g | comm |
| **millimeters** | dimensior |
| **centimeters** | dimensior |
| **kilograms** | weight |
| glass | build |
| gorilla | build |
| plastic | build |
| volume | sound |

You also get the top list of nouns from str_sel_list. This can be used to tweak your mapping list for attributes of categories. Any noun that comes out as the top word but is not in your mapping file should be included. See Listing 3-56 and Figure 3-19.

***Listing 3-56.***

```
###get top words
from collections import Counter
str_sel_list_all = ' '.join(str_sel_list)
str_sel_list_all = str_sel_list_all.replace('.','')
str_sel_list_all_list = Counter(str_sel_list_all.split())
str_sel_list_all_list.most_common()
```

```
[('i', 3812),
 ('phone', 3722),
 ('battery', 486),
 ('screen', 463),
 ('product', 335),
 ('time', 306),
 ('price', 299),
 ('phone,', 287),
```

***Figure 3-19.***

# Step 4: Mapping Each Review to an Attribute

Once you set your mapping file, you will iteratively map each review to an attribute. See Listing 3-57.

***Listing 3-57.***

```
attr_cats = pd.read_csv("attr_cat.csv")

attrs_all = []
for index,row in t3.iterrows():
    ext1 = row["all_attrs_ext"]
    attr_list = []
    for index1,row1 in attr_cats.iterrows():
        wrd = row1["words"]
        cat = row1["cat"]
```

111

```
        if(ext1.find(wrd)>=0):
            attr_list.append(cat)
    attr_list = list(set(attr_list))
    attr_str = ' '.join(attr_list)
    attrs_all.append(attr_str)
```

The attrs_all column in the dataframe contains attributes extracted.

# Step 5: Analyzing Brands

You will now analyze how different brands compare across different categories. For this you chose top brands. You will also classify ratings into a pol_tag. High ratings of 4 and 5 will be classified as positive, low ratings of 1 and 2 as negative, and a rating of 3 as neutral. See Listing 3-58.

**Listing 3-58.** Classifying Ratings

```
t3["attrs"] = attrs_all
t3["pol_tag"] = "neu"
t3.loc[t3.Rating>=4,"pol_tag"] = "pos"
t3.loc[t3.Rating<=2,"pol_tag"] = "neg"
```

The top brands are those that have a minimum of count of 100 reviews in your sample dataset. See Listing 3-59.

**Listing 3-59.** The Top Brands

```
brand_df = pd.DataFrame(t3["Brand Name"].value_counts()).reset_index()
brand_df.columns = ["brand","count"]
brand_df1 = brand_df[brand_df["count"]>=100]
brand_list = list(brand_df1["brand"])

['Samsung',
 'BLU',
 'Apple',
 'LG',
 'Nokia',
 'Motorola',
```

```
'BlackBerry',
'CNPGD',
'HTC']
```

The output of Listing 3-59 is the list of top brands in your dataset. The analysis will be done for the same. You get the most common attributes for positive ratings and the most common attributes for negative ratings with the count using the following function. You create a parameterized variable col3 that contains the ratio of positive to negative score for each attribute. This function will be called at an overall level as well as for each brand. You can now compare the ratios of different brands and get the perceived propositions for each of the brands. See Listing 3-60.

### Listing 3-60.

```
def get_attrs_df(df1,col1,col2,col3):
    list1 = ' '.join(list(df1.loc[(df1.pol_tag=='pos'),"attrs"]))
    list2 = Counter(list1.split())
    df_pos = pd.DataFrame(list2.most_common())

    list1 = ' '.join(list(df1.loc[df1.pol_tag=='neg',"attrs"]))
    list2 = Counter(list1.split())
    df_neg = pd.DataFrame(list2.most_common())

    df_pos.columns = ["attrs",col1]
    df_neg.columns = ["attrs",col2]
    df_all = pd.merge(df_pos,df_neg,on="attrs")
    df_all[col3] = df_all[col1]/df_all[col2]

    return df_all
```

The first df_gen function call here gets the most common positive attributes and the most common negative attributes along with the occurrence in a dataframe. This is then followed by iteratively calling the function for each of the brands in the list of brands. See Listing 3-61 and Figure 3-20.

### Listing 3-61.

```
df_gen = get_attrs_df(t3,"pos_count","neg_count","ratio_all")

for num,i in enumerate(brand_list):
    brand_only_df = t3.loc[t3["Brand Name"]==i]
```

113

```
    col1 = 'pos_count_' + i
    col2 = "neg_count_" + i
    col3 = "ratio_" + i

    df_brand = get_attrs_df(brand_only_df,col1,col2,col3)

    df_gen = pd.merge(df_gen,df_brand[["attrs",col3]],how='left',on="attrs")
df_gen
```

| attrs | ratio_ all | ratio_ Samsung | ratio_ BLU | ratio_ Apple | ratio_ LG | ratio_ Nokia | ratio_ Motorola | ratio_ BlackBerry | ratio_ CNPGD | ratio_ HTC |
|-------|-----------|----------------|------------|--------------|-----------|--------------|-----------------|-------------------|--------------|-----------|
| battery | 1.64 | 2.14 | 1.97 | 1.15 | 1.47 | 1.89 | 1.67 | 0.60 | 1.60 | 1.11 |
| camera | 3.35 | 3.78 | 3.50 | 4.00 | 3.40 | 3.80 | 4.00 | 2.00 | 1.50 | 3.00 |
| display | 1.67 | 2.67 | 2.15 | 0.65 | 2.57 | 1.75 | 2.67 | 0.25 | 0.10 | 0.71 |
| price | 1.06 | 1.13 | 2.21 | 0.64 | 3.00 | 0.86 | 0.67 | 0.67 | 0.36 | 0.11 |
| comm | 1.84 | 4.20 | 1.73 | 1.00 | 3.50 | 3.67 | 4.00 | 1.33 | 0.60 | 2.00 |

***Figure 3-20.***

The first column, ratio_all, is the overall score for the attributes. Overall, people are more favorable on `camera` and least favorable on `price`. In `price`, you can see that the Blu brand seems to be favored better than the rest. In `camera`, Apple and Blackberry stand out. You will plot a chart between top brands and the relative scores on attributes (positive to negative scores). But first you need to prepare the data in a format needed for matplotlib. You first gather all the names of the ratio columns for the top brands. See Listing 3-62 and Figure 3-21.

***Listing 3-62.***

```
ratio_list = []
for i in brand_list:
    ratio_list.append("ratio_" + i)
ratio_list
```

```
['ratio_Apple',
 'ratio_Samsung',
 'ratio_BLU',
 'ratio_LG',
 'ratio_Nokia',
 'ratio_BlackBerry
 'ratio_Motorola',
 'ratio_HTC',
 'ratio_CNPGD']
```

*Figure 3-21.*

Next, you make all of the values in each of the columns as a list of lists. Each set of lists in the list (list_all) corresponds to a brand and the set of values with each list brand corresponds to the attributes of interest. Here you have taken the attributes at the beginning of the dataframe: battery, display, camera, price, comm, and os. See Listing 3-63 and Figure 3-22.

*Listing 3-63.*

```
import matplotlib.pyplot as plt
cols = ["battery","display","camera","price","comm","os"]
list_all = []
for j in ratio_list:

    list1 = df_gen.loc[df_gen.attrs.isin(cols),j].round(2).fillna(0).
    values.flatten().tolist()
    list_all.append(list1)
list_all
```

```
[[2.62, 2.88, 4.12, 3.75, 1.75, 7.5],
 [0.63, 1.78, 2.2, 0.43, 0.2, 0.44],
 [1.52, 2.0, 2.15, 1.71, 2.0, 1.42],
 [1.86, 1.5, 2.4, 1.2, 0.56, 5.0],
 [1.75, 1.25, 2.8, 3.33, 1.2, 3.33],
 [1.75, 1.0, 5.0, 1.33, 0.67, 3.0],
 [1.44, 1.2, 2.2, 1.67, 1.5, 1.5],
 [0.67, 0.33, 0.0, 4.0, 0.62, 1.33],
 [0.5, 1.5, 3.33, 0.75, 0.5, 0.0]]
```

***Figure 3-22.***

Now that the data is prepared, you use matplotlib to plot the charts. You plot each series as a set of X values and increment the list of X values so that the next set moves a little away from the previous set. You have six attributes and you run the loop six times. Each list corresponds to a brand and so the label is for that brand. See Listing 3-64.

***Listing 3-64.***

```python
import matplotlib.pyplot as plt
import numpy as np
X = np.arange(6)

for i in range(6):
    plt.bar(X, list_all[i], width = 0.09,label=brand_list[i])
    plt.xticks(X, cols)

    X=X+0.09

plt.ylabel('Ratio of Positive/Negative')
plt.xlabel('Attributes')
plt.legend(bbox_to_anchor=(1.05, 1), loc='upper left', borderaxespad=0.)
plt.show()
```

The chart in Figure 3-23 is the plot of the same.

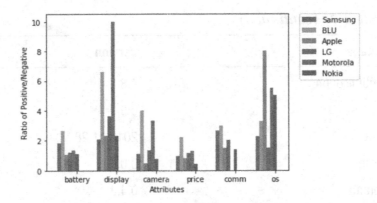

**Figure 3-23.**

As a summary, so far you have explored three methods of sentiment analysis in depth here: lexicons, rules applied with a lexicon, and machine learning analysis. Similar techniques can be applied for emotion or subjectivity/objectivity detection. You have also seen an algorithm for attribute analysis. You can do attribute analysis using supervised learning if you have a corpus trained suitably with different words marked as attributes. Once you have a sentence tagged to a polarity, mood, and attribute, you can derive a lot more insights by slicing and dicing the data with various dimensions.

Table 3-7 lists the packages used in Chapter 3.

**Table 3-7.**  *Packages Used in Chapter 3*

| Package | Version |
| --- | --- |
| azure-cognitiveservices-nspkg | 3.0.1 |
| azure-cognitiveservices-search-nspkg | 3.0.1 |
| azure-cognitiveservices-search-websearch | 1.0.0 |
| azure-common | 1.1.24 |
| azure-nspkg | 3.0.2 |
| backcall | 0.1.0 |
| backports-abc | 0.5 |
| backports.shutil-get-terminal-size | 1.0.0 |

(*continued*)

117

***Table 3-7.*** (*continued*)

| Package | Version |
| --- | --- |
| beautifulsoup4 | 4.8.1 |
| bs4 | 0.0.1 |
| certifi | 2019.11.28 |
| chardet | 3.0.4 |
| colorama | 0.4.3 |
| decorator | 4.4.1d |
| Distance | 0.1.3 |
| enum34 | 1.1.6 |
| futures | 3.3.0 |
| idna | 2.8 |
| ipykernel | 4.10.0 |
| ipython | 5.8.0 |
| ipython-genutils | 0.2.0 |
| isodate | 0.6.0 |
| jedi | 0.13.3 |
| jupyter-client | 5.3.4 |
| jupyter-core | 4.6.1 |
| msrest | 0.6.10 |
| nltk | 3.4.3 |
| numpy | 1.16.6 |
| oauthlib | 3.1.0 |
| pandas | 0.24.2 |
| parso | 0.4.0 |
| pathlib2 | 2.3.5 |
| pickleshare | 0.7.5 |

(*continued*)

***Table 3-7.*** (*continued*)

| Package | Version |
| --- | --- |
| pip | 19.3.1 |
| prompt-toolkit | 1.0.15 |
| pybind11 | 2.4.3 |
| pydot-ng | 1.0.1.dev0p |
| Pygments | 2.4.2 |
| pyparsing | 2.4.6 |
| python-dateutil | 2.8.1 |
| pywin32 | 227 |
| pyzmq | 18.1.0 |
| requests | 2.22.0 |
| requests-oauthlib | 1.3.0 |
| scandir | 1.10.0 |
| setuptools | 44.0.0.post20200106 |
| simplegeneric | 0.8.1 |
| singledispatch | 3.4.0.3 |
| six | 1.13.0 |
| sklearn | 0 |
| soupsieve | 1.9.5 |
| speechrecognition | 3.8.1 |
| tornado | 5.1.1 |
| traitlets | 4.3.3 |
| urllib3 | 1.25.7 |
| wcwidth | 0.1.8 |
| wheel | 0.33.6 |
| win-unicode-console | 0.5 |
| wincertstore | 0.2 |

# CHAPTER 4

# NLP in Banking, Financial Services, and Insurance (BFSI)

The banking and financial industries have been making data-driven decisions for more than a century. Since a wrong decision could have a heavy cost for a financial institution, they have been one of the early adopters of big data. A lot of machine learning use cases in the banking, financial services, and insurance industries (BFSI) have been using structured data like transactional history or CRM history. However, over the last few years there has been an increasing tendency to use text data to mainstream underwriting and risk or fraud detecting.

## NLP in Fraud

One of the challenges in banking, especially commercial lending, is to understand the risks of organizations. When the transaction history of the entities is not available or not sufficient, financial institutions can look at publicly available data like news or online blogs to understand any mention of the entities. There are companies that supply such data to lending institutions. Our first problem is to find out the mentions of entities in a news corpus and then understand if they are mentioned in a positive or negative light.

You will be using a dataset from Kaggle (`www.kaggle.com/sunnysai12345/news-summary`) for this exercise. The news articles in this dataset have been scraped from various newspapers. The summary for the news articles have been scraped from a mobile app company InShorts. Your objective here is to identify mentions of company names and the sentiment associated with the mentions. See Listing 4-1 and Figure 4-1. Please note that all packages used in this chapter are listed in Table 4-6.

121

© Mathangi Sri 2021
M. Sri, *Practical Natural Language Processing with Python*, https://doi.org/10.1007/978-1-4842-6246-7_4

***Listing 4-1.***  Importing the Sample Dataset

```
import pandas as pd
import nltk
t1 = pd.read_csv("news_summary.csv",encoding="latin1")
t1.head()
```

| | author | date | headlines | read_more | text | ctext |
|---|---|---|---|---|---|---|
| 0 | Chhavi Tyagi | 03 Aug 2017,Thursday | Daman & Diu revokes mandatory Rakshabandhan in... | http://www.hindustantimes.com/india-news/raksh... | The Administration of Union Territory Daman an.. | The Daman and Diu administration on Wednesday .. |
| 1 | Daisy Mowke | 03 Aug 2017,Thursday | Malaika slams user who trolled her for 'divorc... | http://www.hindustantimes.com/bollywood/malaik... | Malaika Arora slammed an Instagram user who tr... | From her special numbers to TV?appearances, Bo... |
| 2 | Arshiya Chopra | 03 Aug 2017,Thursday | 'Virgin' now corrected to 'Unmarried' in IGIMS... | http://www.hindustantimes.com/patna/bihar-igim... | The Indira Gandhi Institute of Medical Science... | The Indira Gandhi Institute of Medical Science... |
| 3 | Sumedha Sehra | 03 Aug 2017,Thursday | Aaj aapne pakad liya: LeT man Dujana before be... | http://indiatoday.intoday.in/story/abu-dujana- | Lashkar-e-Taiba's Kashmir commander Abu Dujana... | Lashkar-e-Taiba's Kashmir commander Abu Dujana... |
| 4 | Aarushi Maheshwari | 03 Aug 2017,Thursday | Hotel staff to get training to spot signs of s... | http://indiatoday.intoday.in/story/sex-traffic... | Hotels in Maharashtra will train their staff t... | Hotels in Mumbai and other Indian cities are t... |

***Figure 4-1.***  *Sample dataset from Kaggle*

The columns that are of importance here are headlines, text (this is the summary), and ctext (the full text). You will try to extract organizations from any of these three columns. In order to do so, you will concatenate all the columns into a single column. See Listing 4-2.

***Listing 4-2.***  Concatenating Columns

```
t1["imp_col"] = t1["headlines"] + " " + t1["text"] + " " + t1["ctext"]
```

You want to extract the named entities of the given text. Recall from the last chapter that named entities are any person, place, organization, etc. mentioned in a text corpus. In your case, you are interested in extracting organization names from the text corpus.

# Method 1: Using Existing Libraries

You can use existing libraries to extract named entities. Listing 4-3 is an example of using the NLTK library for extracting named entities.

***Listing 4-3.*** The NLTK Library

```
###Example of NER
sent = "I work at Apple and before I used to work at Walmart"
tkn = nltk.word_tokenize(sent)
ne_tree = nltk.ne_chunk(nltk.pos_tag(tkn))
print(ne_tree)

(S
  I/PRP
  work/VBP
  at/IN
  (ORGANIZATION Apple/NNP)
  and/CC
  before/IN
  I/PRP
  used/VBD
  to/TO
  work/VB
  at/IN
  (ORGANIZATION Walmart/NNP))
```

You can apply the library to a sample of sentences and check if it is efficient and able to capture all of the names. The returned Named Entity tree is parsed by the function get_chunk. You retrieve all words that are labelled as "organization." See Listings 4-4 and 4-5.

***Listing 4-4.***

```
samp_sents = t1.loc[0:10,"imp_col"]

defget_chunk(ne_tree):
    org_list = []
for chunk in ne_tree:
if hasattr(chunk, 'label'):
            str1 = chunk.label()
if(str1.find("ORG")>=0):
                str2 = ' '.join(c[0] for c in chunk)
                org_list.append(str2)
return org_list
```

*Listing 4-5.*

```
org_list_all = []
for i in samp_sents:
    tkn = nltk.word_tokenize(i)
    ne_tree = nltk.ne_chunk(nltk.pos_tag(tkn))
    org_list = get_chunk(ne_tree)
    org_list_all.append(org_list)
org_list_all
```

Let's examine org_list_all and see if the organizations are captured. A sample is shown in Listing 4-6.

*Listing 4-6.*

```
[['Diu',
'Administration of Union',
'Daman',
'Daman',
'UT',
'Diu',
'Gujarat',
'BJP',
'Centre',
'RSS',
'RSS'],
 ['Hindi', 'Munni Badnam'],
 ['IGIMS',
'Indira Gandhi Institute',
'Medical Sciences',
'IGIMS',
'Marital',
'Indira Gandhi Institute',
'Medical Sciences',
'IGIMS',
```

```
'All India Institute',
'Medical Sciences',
'HT Photo'],
```

As you can see, NLTK has done a decent job of retrieving organization names from the news corpus.

## Method 2: Extracting Noun Phrases

Another way to retrieve names of organizations is to extract noun phrases from a corpus. Organizations or any other "named entities" are nouns or noun phrases. You will see a lot more names than you saw in the last example since not all nouns are names of organizations. You will use a library called Spacy for the exercise. https://spacy.io/ is an open source natural language toolkit. Spacy can help in a host of NLP functions including parts-of-speech tagging and named entity recognition. You are going to do a deep dive into the parts-of-speech functionality of Spacy. Parts-of-speech parsing is a technique where a sentence is divided into its grammatical components and the underlying structure. Let's see a small example of extracting parts of speech using Spacy. You can install Spacy by using pip install. See Listing 4-7.

***Listing 4-7.*** Installing Spacy

```
!pip install spacy
```

You need Spacy's models to perform NLP tasks. You download the en_core_web_sm library by using the code in Listing 4-8. More details of Spacy English models are available at https://spacy.io/models/en. Also see Listing 4-9.

***Listing 4-8.*** Downloading a Library

```
!python -m spacy download en_core_web_sm
```

***Listing 4-9.*** Importing Spacy

```
import spacy
nlp = spacy.load("en_core_web_sm")
doc = nlp("This is to explain parts of speech in Spacy")
for i in doc:
print (i,i.pos_)
```

```
This DET
is VERB
to PART
explain VERB
parts NOUN
of ADP
speech NOUN
in ADP
Spacy PROPN
```

Each word gets assigned a parts-of-speech tag. Sample POS tags and their meanings are listed at https://spacy.io/api/annotation. For instance, DET is a determiner and refers to *a, an,* or *the*. ADP is an adpositional phrase and can refer to *to, in, during,* etc. You are more interested in nouns for the purpose at hand. Let's use Spacy to extract nouns from the sample sentences and compare with the earlier results from method 1. See Listing 4-10.

***Listing 4-10.*** Extracting Nouns

```
np_all = []
for i in samp_sents:
    doc = nlp(i)
for np in doc.noun_chunks:
        np_all.append(np)
np_all
[Daman,
 Diu,
 mandatory Rakshabandhan,
 The Administration,
 Union Territory Daman,
 Diu,
 its order,
 it,
 women,
 rakhis,
 their male colleagues,
 the occasion,
```

The output from Listing 4-10 is just a Start from here sample and you can see that there are many more words that got extracted here than from method 1. Now, check if there is any pattern of organization words that method 1 missed. Compare the output of method 1's org_list_all and np_all. This could give you some pointers on what type of words method 1 missed. See Listing 4-11.

***Listing 4-11.***

```
np_all_uq = set(np_all)
merged = set(merged)
np_all_uq - merged

{(DGCA,
 training,
 an alert,
 helidecks,
 The actor,
 The woman,
 Lucknow,
 Dujana,
 about a million small hotels,
 most cases,
 This extraordinary development,
 bachelors,
 the bond,
 forced labor,
 FSSAI,
```

One observation is that method 1 missed acronyms like DGCA and FSSAI. These types of acronyms could be recognized and appended to the first output. In cases where the list of companies is known, you could extract these names and compare them with the list at hand by using a suitable "fuzzy" or "text similarity" operations.

# Method 3: Training Your Own Model

Method 1 and Method 2 are based on libraries that are on pretrained models. Another approach is to train your own model to detect named entities. The concept behind the named entities model is that the type of entity of an word can be predicted based on the words surrounding the word and the word itself. I have taken the example of the sentence "I work at Apple and before I used to work at Walmart." In this case, the data structure is something like that shown in Table 4-1.

*Table 4-1.* *Data Structure*

| Word | prev_word | next_word | pos_tag | prev_pos_tag | next_pos_tag | Org |
|------|-----------|-----------|---------|--------------|--------------|-----|
| I | beg | work | PRP | beg | PRP | 0 |
| work | I | at | VBP | PRP | VBP | 0 |
| at | work | Apple | IN | VBP | IN | 0 |
| Apple | at | and | NNP | IN | NNP | 1 |
| and | Apple | before | CC | NNP | CC | 0 |
| before | and | I | IN | CC | IN | 0 |
| I | before | used | PRP | IN | PRP | 0 |
| used | I | to | VBD | PRP | VBD | 0 |
| to | used | work | TO | VBD | TO | 0 |
| work | to | at | VB | TO | VB | 0 |
| at | work | Walmart | IN | VB | IN | 0 |
| Walmart | at | end | NNP | IN | end | 1 |

The last column, org, is manually labelled as to whether the word is an organization or not. With the word-level features and using the manually labeled data as the expected output, you can custom train a supervised classifier.

For your exercise, you will use the named entity recognizer dataset from Kaggle. Details of the dataset can be found at www.kaggle.com/abhinavwalia95/entity-annotated-corpus. Essentially it's a list of sentences broken down into a word level and manually labelled at the word level into Organization, Person, Time, Geo-political entity, etc. Table 4-2 shows a sample of the dataset.

*Table 4-2. NER Training Dataset*

| | lemma | next-lemma | next-next-lemma | next-next-pos | next-next-shape | next-next-word | next-pos | next-shape | next-word | word | tag |
|---|---|---|---|---|---|---|---|---|---|---|---|
| 0 | thousand | of | demonstr | NNS | lowercase | demonstrators | IN | lowercase | of | Thousands | O |
| 1 | of | demonstr | have | VBP | lowercase | have | NNS | lowercase | demonstrators | of | O |
| 2 | demonstr | have | march | VBN | lowercase | marched | VBP | lowercase | have | demonstrators | O |
| 3 | have | march | through | IN | lowercase | through | VBN | lowercase | marched | have | O |
| 4 | march | through | london | NNP | capitalized | London | IN | lowercase | through | marched | O |
| 5 | through | london | to | TO | lowercase | to | NNP | capitalized | London | through | B-geo |
| 6 | london | to | protest | VB | lowercase | protest | TO | lowercase | to | London | O |
| 7 | to | protest | the | DT | lowercase | the | VB | lowercase | protest | to | O |
| 8 | protest | the | war | NN | lowercase | war | DT | lowercase | the | protest | O |
| 9 | the | war | in | IN | lowercase | in | NN | lowercase | war | the | O |

Not all columns are shown but you get the idea. You will be using a supervised classifier for this approach, so you need to treat the high-dimensional categorical features herewith. You could concatenate all the word-level features into a single column and apply a tf-idf vectorizer with unigram as a parameter. Please note that the order of these categorical features does not make a difference, so you could treat the feature set as one sentence where the sequences do not matter.

In this method of modeling, the named entity recognizer depends on the significance of a given word and does not generalize the context of the word. For instance, in this method of modeling, the named entity recognizer follows the significance of a given word but will not generalize to the context. For example, in the above sentence, we could replace the word "before" with "earlier" and the meaning would not change. See Figure 4-2.

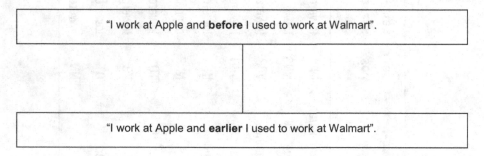

*Figure 4-2.*

The other problem with this method is the sparsity of the dataset given the vocabulary of the words could run into tens of thousands if not millions. You need a dense vector to represent the word features in your table. Word embeddings can help us overcome these disadvantages and is explained in the section below.

## Word Embeddings

The bag-of-words method described in Chapter 1 suffers from the disadvantage of not being able to learn context. Word embeddings is a technique that helps learn context and shifts the NLP paradigm from a syntactic approach to a semantic approach. Consider search engines. When users search, for instance, for domestic animals, they expect results on cats, dogs, etc. Observe that the words are not lexically similar but they are absolutely related. Word embeddings help associate words to context.

The intuition behind word embeddings is that you can learn the context of the word by the words surrounding it. For example, in Figure 4-3 all the highlighted words appear in a similar context of learning. The words surrounding a given word give it its meaning. Hence from a large corpus, by comparing words surrounding a given word we can arrive at the semantics of the word.

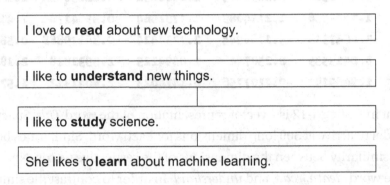

*Figure 4-3.* *Example sentences for word embedding*

Around 2013, neural networks were used to learn the embeddings of words. Using a technique called word2vec, you could represent a single word into N dimensions. Once you get a word represented in N dimensions, you can arrive at the cosine similarity of the dimensions to understand the similarity between the words. There are also pretrained libraries available to get the dimensions for each word. Let's see an example from Spacy. See Listing 4-12.

*Listing 4-12.* Spacy

```
nlp = spacy.load("en_core_web_sm")
doc = nlp("This is to explain parts of speech in Spacy")
for i in doc:
    print(i.text, i.vector)
```

```
This [-2.2498417  -3.3314195  -5.568074    3.67963    -4.612324   -0.48435217
  0.27712554 -0.77937424 -0.8964542  -1.0418985  -0.7162781   2.076798
 -3.6910486   2.2700787  -3.4369807  -1.8229328   0.1314069   0.22531027
  5.15489    -0.49419683  1.0052757  -1.8017105   2.1641645   2.0517178
 -2.1288662   3.7649343  -0.66652143 -2.9879086   0.05217028  4.8123035
  4.051547    0.98279643 -2.6945567  -3.4998686  -1.7214956  -0.61138093
```

```
 1.4845726    2.335553    -3.3347645  -0.62267256 -2.6027427   0.5974684
-4.0274143    3.0827138   -2.8701754  -2.8391452   6.236176     2.1810527
 3.2634978   -3.845146    -1.7427144   1.1416728  -3.019658    -0.505347
-0.564788     0.8908412   -1.5224171  -0.8401189  -2.4254866    2.3174202
-4.468281    -2.662314    -0.29495603  3.1088934  -0.69663733  -1.3601663
 3.0072615   -5.8090506   -1.2156953  -2.2680192   1.2263682   -2.620054
 4.355402     1.1643956    1.2154089   1.1792964  -0.9964176    1.4115602
 0.2106615   -3.0647554   -5.1766944   2.7233424  -0.36179888  -0.58938277
 3.1080005    8.0415535    8.250744   -1.0494225  -2.9932828    2.1838582
 3.6895149    1.9469192   -0.7207756   6.1474023   2.4041312    1.5213318 ]
```

The output in Listing 4-12 is a vector representation of the word *This*. Spacy trained a word2vec model to arrive at 300 long dimensions for each word. Similarities between the words are the similarity between the dimensions of the respective words. Now let's take the example of the words *learn, read,* and *understand*. In order to contrast the similarities, you will also take the context vector of the word *run*. You are experimenting with another Spacy model, en_core_web_md. First, download the model and then use it in your script. See Listings 4-13 and 4-14.

***Listing 4-13.*** Downloading Spacy

```
!python -m spacy download en_core_web_md
```

***Listing 4-14.*** Using Spacy

```
import spacy
nlp = spacy.load("en_core_web_md")
doc1 = nlp(u'learn')
doc2 = nlp(u'read')
doc3 = nlp(u'understand')
doc4 = nlp(u'run')
for i in [doc1,doc2,doc3,doc4]:
    for j in [doc1,doc2,doc3,doc4]:

        if((i.vector_norm>0) & (j.vector_norm>0)):
            print (i,j,i.has_vector,j.has_vector,i.similarity(j))
learn learn True True 1.0
learn read True True 0.47037032349231994
```

```
learn understand True True 0.7617364524642122
learn run True True 0.32249343409210124
read learn True True 0.47037032349231994
read read True True 1.0
read understand True True 0.5366261494062171
read run True True 0.2904269408366301
understand learn True True 0.7617364524642122
understand read True True 0.5366261494062171
understand understand True True 1.0
understand run True True 0.32817161669079264
run learn True True 0.32249343409210124
run read True True 0.2904269408366301
run understand True True 0.32817161669079264
run run True True 1.0
```

As you can see here, the similarity between *learn* and *understand* is much stronger compared to *run* and *understand*.

## Word2Vec

Word2Vec is one technique of creating word embeddings. Here you use words that are within a window size as the "independent" variables to predict the target word. The input and output data for the sentence "I like to read about technology" are shown in Table 4-3.

*Table 4-3.*

| Input | Output |
|---|---|
| I | Like |
| Like | I |
| Like | To |
| To | Like |
| To | I |
| Read | To |
| Read | About |
| About | Read |
| About | technology |
| technology | About |

A shallow, single layer softmax model is trained and the weight matrix of the hidden layer provides the dimensions of the words. A very good explanation of how word2vec is trained is explained at www.analyticsvidhya.com/blog/2017/06/word-embeddings-count-word2veec/. The architecture of word2vec is provided in Figure 4-4 and Figure 4-5.

# CBOW

With CBOW, each word in the sentence becomes the target word. The input words are the words surrounding it with a window size. The below example is with window size = 2. The architecture of CBOW is shown in Figure 4-4.

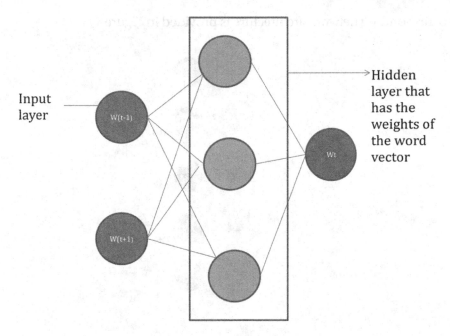

*Figure 4-4.*

The w(t-1) and w(t+1) are one hot encoded vectors. The size of each is the size of the vocabulary. The same can be said of the output vector and it is as long as the vocabulary. The center layer coded in green (the hidden layer) contains the weight matrix. You can average or sum the weight matrix corresponding to each of the input vectors.

Another way to build the word2vec is to turn the CBOW approach on its head and get a skip gram model. Here you assume that a single word in a sentence should be picking up the surrounding words. An example of the input and output representation is provided in Table 4-4 with a window size of 2.

*Table 4-4.*

| Input | Word_output_t-2 | Word_output_t-1 | Word_output_t+1 | Word_output_t+2 |
|-------|-----------------|-----------------|-----------------|-----------------|
| I | | | like | to |
| like | | I | to | read |
| to | I | like | read | about |
| read | like | to | about | technology |
| about | to | read | read | |
| technology | read | about | | |

The corresponding network architecture is provided in Figure 4-5.

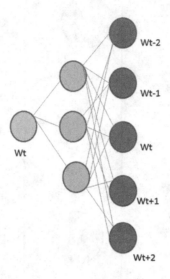

*Figure 4-5.*

The hidden layer or the word vectors in this case is 3. So you input a single word in a sentence and train the model to predict words surrounding the sentences. In Figure 4-5, each of the nodes is a one hot vector and this is a large output to predict if the vocabulary sizes are large. In order to circumvent the problem, a technique called negative sampling is used. Instead of looking at the problem as a multi-class classifier, the input word and the context words are treated as inputs and the output is whether the word combination is probable or not. This converts the problem to a binary classifier. Along with probable sets of words together, improbable sets are chosen from the corpus and they are labelled as the negative set (the co-occurring words as the positive set). The negative set is chosen from a separate probability distribution. Details can be found at https://papers.nips. cc/paper/5021-distributed-representations-of-words-and-phrases-and-their- compositionality.pdf.

You will use your initial corpus of the news dataset t1 and train a word2vec using the gensim library. Before you apply the gensim library to your corpus, you need to preprocess the documents. You will remove stop words from the corpus and "stem" the words. Stemming is a process of bringing words back to their root forms and hence similar words can be normalized. You will need to install the gensim package using pip install. See Listings 4-15 and 4-16.

***Listing 4-15.*** Installing gensim

```
!pip install gensim
```

***Listing 4-16.*** Using gensim

```
from gensim.parsing.porter import PorterStemmer
p = PorterStemmer()
p.stem_documents(["He was dining", "We dined yesterday"])
```

```
['he wa dine', 'we dine yesterdai']
```

The words *dining* and *dine* got replaced with the same word, *dine*, in the Listing 4-16 example. This can help normalize words to their roots. You will continue with your preprocessing steps. You first remove stop words with the pandas string `replace` method and the stop_words library. This is followed by stemming the words. Before you proceed, you must first install the stop_words library using pip install. See Listings 4-17 and 4-18.

***Listing 4-17.*** Installing stop_words

```
!pip install stop_words
```

***Listing 4-18.***

```
import stop_words
eng_words = stop_words.get_stop_words('en')

for i in eng_words:
    if(len(i)>1):
        wrd_repl = r"\b" + i + r"\b"

        t1["h1_stop"] = t1["h1_stop"].str.replace(wrd_repl,"")
t1["h2_stop"] = pd.Series(p.stem_documents(t1["h1_stop"]))
```

Word2Vec in gensim takes list of list words as input. You convert the headlines column into a list of lists. It's better to use ctext as it contains a huge corpus of words, but I used the headlines column for an example. See Listing 4-19.

*Listing 4-19.*

```
sentences = list(t1["h2_stop"].str.split())
Sentences[0:10]
[['daman',
  '&',
  'diu',
  'revok',
  'mandatori',
  'rakshabandhan',
  'offic',
  'order'],
 ['malaika', 'slam', 'user', 'troll', "'divorc", 'rich', "man'"],
 ["'virgin'", 'now', 'correct', "'unmarried'", "igims'", 'form'],
 ['aaj', 'aapn', 'pakad', 'liya:', 'let', 'man', 'dujana', 'kill'],
```

Now let's build the word2vec gensim model. The `size` option is the number of dimensions you want the word to be represented in. `min_count` is the minimum number of documents the word should be in. The `iter` option is the number of epochs of the neural network. See Listing 4-20.

*Listing 4-20.*

```
model = gensim.models.Word2Vec(iter=10,min_count=10,size=200)
model.build_vocab(sentences)
token_count = sum([len(sentence) for sentence in sentences])
model.train(sentences,total_examples = token_count,epochs = model.iter)
model.wv.vocab
{'offic': <gensim.models.keyedvectors.Vocab at 0xf745320>,
 'order': <gensim.models.keyedvectors.Vocab at 0x1aab8780>,
 'slam': <gensim.models.keyedvectors.Vocab at 0x1aab8e80>,
 'user': <gensim.models.keyedvectors.Vocab at 0x1aab8a20>,
 'now': <gensim.models.keyedvectors.Vocab at 0x1aaa80f0>,
 'form': <gensim.models.keyedvectors.Vocab at 0x1aaa8320>,
 'let': <gensim.models.keyedvectors.Vocab at 0x1aaa8908>,
 'man': <gensim.models.keyedvectors.Vocab at 0x1325a160>,
```

```
'kill': <gensim.models.keyedvectors.Vocab at 0x15bd87f0>,
'hotel': <gensim.models.keyedvectors.Vocab at 0x5726198>,
'staff': <gensim.models.keyedvectors.Vocab at 0x5726f98>,
'get': <gensim.models.keyedvectors.Vocab at 0x13275518>,
'train': <gensim.models.keyedvectors.Vocab at 0x13275240>,
```

Each of the words in the Listing 4-20 example has a 200-dimensional vector and has a separate space allotted to store the dimensions of the word. The Listing 4-20 output displays a sample of the vocabulary list. You can quickly check if the model is doing some justice to semantic similarity. gensim has a function named most_similar that prints out the closest word for the given word. Basically, the 200-dimensional vector of the target word is compared with all the other words in the corpus and the most similar are printed in order. See Listing 4-21.

***Listing 4-21.***

```
model.most_similar('polic')
[('kill', 0.9998772144317627),
 ('cop', 0.9998742341995239),
 ('airport', 0.9998739957809448),
 ('arrest', 0.9998728036880493),
 ('new', 0.9998713731765747),
 ('cr', 0.9998686909675598),
 ('indian', 0.9998684525489807),
 ('govt', 0.9998669624328613),
 ('dai', 0.9998668432235718),
 ('us', 0.9998658895492554)]
```

For the stemmed word *polic,* you can see that *cop, arrest,* and *kill* are semantically similar. The similarity can be made better by using the complete text rather than the headlines, like you considered. The word embeddings for the word can be retrieved using the code in Listing 4-22.

***Listing 4-22.***

```
model["polic"]
array([ 0.04353377, -0.15353541,  0.01236599,  0.17542918, -0.02843889,
        0.0871359 , -0.1178024 , -0.00746543, -0.03241869,  0.0475546 ,
        0.04885347, -0.05488129, -0.08609696,  0.15661193,  0.1471964 ,
        0.002795,    0.06438455, -0.12603344,  0.00585101,  0.10587147,
        0.03390319,  0.35148793, -0.06524974,  0.07119066,  0.17404315,
        0.02006942, -0.1783511 ,  0.02980262,  0.26949257, -0.07674567,
```

Listing 4-22 is a sample of the 200-dimension vector for the word *polic*.

## Other word2vec Libraries

Another popular library for word2vec is fasttext. This is a library released by Facebook. The primary difference is that the input vector being considered is a set of ngrams of the word. For example, *transform* gets broken down into *tra, ran, ans,* etc. This helps the word2vec model to provide vector representation for most words because they will be broken down into ngrams and their weights recombined to give a final output. You will see an example.

You will use your preprocessed column h1_stop t1 dataset example. You will replace all numbers and remove all words with a length of three from the column. See Listing 4-23.

***Listing 4-23.***

```
t1["h1_stop"] = t1["h1_stop"].str.replace('[0-9]+','numbrepl')
t1["h1_stop"] = t1["h1_stop"].str.replace('[^a-z\s]+','')
list2=[]
for i in t1["h1_stop"]:
  list1 = i.split()
  str1 = ""
for j in list1:
if(len(j)>=3):
      str1 = str1 + " " + j
  list2.append(str1)
t1["h1_proc"] = list2
```

Now you extract word tokens from the h1_proc column. You have used word_ punctuation_tokenizer to tokenize words. The punctuation tokenizer treats the punctuation as separate tokens compared to the tokenize function from NLTK. See Listings 4-24 through 4-26.

***Listing 4-24.***

```
word_punctuation_tokenizer = nltk.WordPunctTokenizer()
word_tokenized_corpus = [word_punctuation_tokenizer.tokenize(sent)
for sent in t1["h1_proc"]]
```

***Listing 4-25.***

```
from gensim.models.fasttext import FastText as FT_gen
model_ft = FT_gen(word_tokenized_corpus,size=100,
                  window=5,
                  min_count=10,

                  sg=1,
                  iter=100
                  )
```

***Listing 4-26.***

```
print(model_ft.wv['police'])
```

```
[-0.29888302 -0.57665455  0.08610394 -0.3632993  -0.11655446  0.77178127
  0.00754398  0.326833    0.6293257  -0.6018732   0.18104246  0.14998026
 -0.25582018 -0.10343572  0.26390004 -0.40456587 -0.13059254  0.7299065
  0.56476426 -0.34973195  0.2719949  -0.23201875 -0.5852624  -0.233576
  0.79868716  0.13634606  0.34833038  0.09759299  0.03199455  0.31180373
 -0.27072772 -0.18527883 -0.58960485  0.01390363  0.50662494  0.65151244
 -0.47332478  0.03739219  0.4098077   0.41875887 -0.4502848   0.41489652
  0.13763584  0.6388822  -0.7644631   0.02981189  0.7131902   0.13380833
 -0.3466939   0.5062037   0.01952425 -0.14834954 -0.29615352  0.11298565
  0.01239631  0.22170404  0.87028074  0.30836678  0.171935    0.06707167
  0.6141744   0.7583458   0.8327537  -0.9569695  -0.5542731  -1.0179517
 -0.5757873  -0.2523322  -0.64286023  0.5246012  -0.00269936 -0.11349581
```

```
 -0.34673667  0.13290115   0.3713985  -0.10439423 -0.2525865    0.1401525
 -0.39776105  0.43563667   0.50854915 -0.32810602  0.5654142   -0.60424364
  0.14617543 -0.03651292   0.01708415 -0.16520758 -0.89801085  -0.7447387
 -0.47594827  0.2536381   -0.5689429  -0.46556026  0.2765042   -0.35487294
  0.37178138 -0.12994497   0.17699395  0.79665047]
```

The parameters window size represents the window of the words to train on, dimensions is the total number of words in the dimensions, and minimum count is the minimum number of documents that the words should be part of for it to be considered for the analysis. Now, try to get the dimensions of a word that is not in the initial vocabulary. Let's consider the word *administration*, for example. See Listing 4-27.

***Listing 4-27.***

```
print ('administration'in model_ft.wv.vocab)
model_ft["administration"]
```

**False**
```
array([-0.45738608,  0.01635892,  0.4138502 , -0.39381024, -0.03351381,
        0.5652306 ,  0.15023482, -0.06436284, -0.30598623,  0.22672811,
       -0.0371761 ,  0.55359524, -0.30920455, -0.09809569,  0.3087636 ,
        0.14136946, -1.0166188 , -0.18632148,  0.54922   , -0.13646998,
       -0.26896936, -0.45346558, -0.01610097, -0.03324416, -0.5120829 ,
        0.29224998,  0.14098023, -0.29888844, -0.03517408, -0.44184253,
        0.00370641,  0.51012856, -0.6507577 , -0.5389353 ,  0.49002212,
        0.37082076, -0.45470816,  0.0273569 ,  0.15264036,  0.16321276,
       -0.15225472,  0.5453094 , -0.31627965, -0.37927952, -0.13029763,
        0.23256542,  0.6777767 , -0.09562921, -0.27754757,  0.45492762,
        0.21526024,  0.6568411 , -0.14619622,  0.9491824 ,  0.5132987 ,
        0.5608091 ,  0.3351915 , -0.19633797,  0.17734666, -1.2677175 ,
        0.70604306, -0.120744  ,  0.19599114, -0.06684009, -0.09474371,
       -0.06331848, -0.5224171 , -0.17912938, -0.9254168 , -0.44491112,
       -0.20046434,  0.39295596, -0.14251098,  0.01547457, -0.0033557 ,
```

As you can see in Listing 4-27, *administration* was not in the initial vocabulary but you were able to get the vector for *administration* since fasttext can internally combine the trigrams of the word.

# Applying Word Embeddings to Supervised Learning

Going back to where you started to build a custom classifier for NER, you need a dense vector space to represent the words in the table to input into the algorithm. There are two ways to use word embeddings in the supervised method.

## Method 3 – Approach 1

You can use the embeddings for each of the words generated from gensim in your supervised model (word embeddings approach 1). In order to use this 200-dimensional vector in your model, you can use it as 200 columns or this vector can be summed or averaged. Another method is to multiply the tf-idf vector value of the word with the sum or average of the vector values.

## Method 3 – Approach 2

Since you are generating the embeddings for the supervised problem of classifying a given word as an organization or not, you can use embeddings better. You can directly train a neural network and learn the embeddings as part of the training process. You can make use of Keras to learn the word embeddings as a layer in the supervised model (word embeddings approach 2). Keras is an open source neural network library in Python (https://keras.io/). Figures 4-6 and 4-7 explain the two approaches of using embeddings in a supervised model.

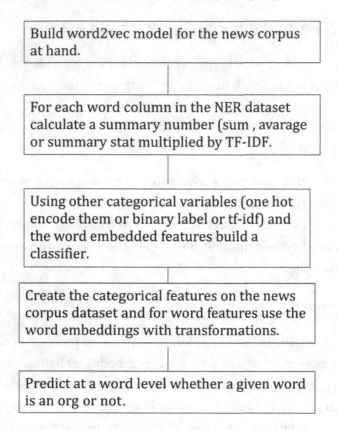

*Figure 4-6.* *Word embeddings approach 1*

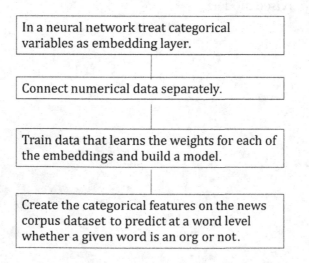

*Figure 4-7.* *Word embeddings using approach 2*

As explained in Figure 4-7, for word embeddings using approach 2, in order to train a classifier for NER, you need to look at a dataset where word-level tags are available. You will use the Kaggle dataset in Table 4-1 for the same. Once you train the model, you will apply the model to your news corpus dataset.

The Keras "embedding layer" learns the embedding vector as a process of training itself. Unlike the word2vec approach, there is no concept of window size here. You first need to label your categorical features, mention the embedding size for each of the categorical variables, flatten it, and then connect to dense layers. The architecture of the network is represented in Figure 4-8.

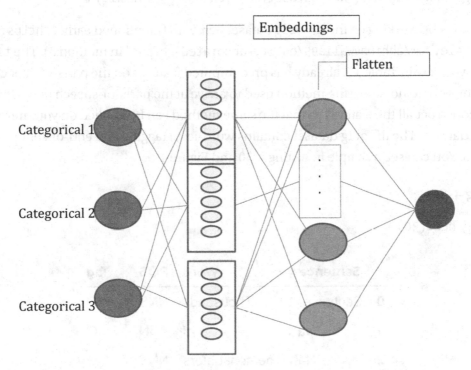

***Figure 4-8.*** *Categorical embeddings*

The input variable here is the set of categorical variables. Then you have an embedding layer that embeds the categorical values into N dimensions and then you flatten and connect to a dense layer and treat the classifier as a classic binary cross-entropy model. You now start implementing the same. See Listings 4-28 and 4-29.

***Listing 4-28.***

```
import pandas as pd
from nltk.stem import WordNetLemmatizer
import numpy as np
import nltk
```

***Listing 4-29.***

```
t1 = pd.read_csv("ner.csv",encoding='latin1', error_bad_lines=False)
df_orig = pd.read_csv("ner_dataset.csv",encoding='latin1')
```

You will be working on the df_orig dataset herewith (mentioned earlier "https://www.kaggle.com/abhinavwalia95/entity-annotated-corpus" in method 3). The t1 dataset you saw in Table 4-1 already has precomputed features on the parts-of-speech tags. However, you do not know the method used to arrive at the parts-of-speech tags. Hence you reconstruct all the features so that the same method can be applied on your news corpus dataset. The df_orig dataset contains word-level tagging and end-of-sentence markers. You can see a sample in Listing 4-30 and Figure 4-9.

***Listing 4-30.***

```
df_orig.head(30)
```

|   | Sentence # | Word | POS | Tag |
|---|---|---|---|---|
| 0 | Sentence: 1 | Thousands | NNS | O |
| 1 | NaN | of | IN | O |
| 2 | NaN | demonstrators | NNS | O |
| 3 | NaN | have | VBP | O |
| 4 | NaN | marched | VBN | O |
| 5 | NaN | through | IN | O |
| 6 | NaN | London | NNP | B-geo |
| 7 | NaN | to | TO | O |

***Figure 4-9.***

146

Sentence # is the sentence marker and Tag is the manual tag. Ignore the POS tag and reconstruct the sentence and use the NLTK library to identify the POS tag. The reason you do this is to standardize themethodology between training data and inferencing data. Once you get the sentence and identify the POS tag, you can generate the "prev" and "next" features. See Listing 4-31.

*Listing 4-31.*

```
##cleaning the columns names and filling missing values
df_orig.columns = ["sentence_id","word","pos","tag"]
df_orig["sentence_id"] = df_orig["sentence_id"].fillna('none')
df_orig["tag"] = df_orig["tag"].fillna('none')
```

A quick glance at the tags distribution herewith is shown in Listing 4-32 and Figure 4-10.

*Listing 4-32.*

```
df_orig["tag"].value_counts()
```

```
                    O        887908
                    B-geo     37644
                    B-tim     20333
                    B-org     20143
                    I-per     17251
                    B-per     16990
                    I-org     16784
                    B-gpe     15870
                    I-geo      7414
                    I-tim      6528
                    B-art       402
                    B-eve       308
                    I-art       297
                    I-eve       253
                    B-nat       201
                    I-gpe       198
                    I-nat        51
                    Name: tag, dtype: int64
```

*Figure 4-10.*

B and I represent the beginning or intermediate. You are interested in the org tags wherever they are present. Listing 4-33 reconstructs the sentence in df_orig and gets the POS tags of the word-level features.

***Listing 4-33.***

```
lemmatizer = WordNetLemmatizer()

list_sent = []
word_list = []
tag_list = []
for ind,row in df_orig[0:100000].iterrows():
    sid = row["sentence_id"]
    word = row["word"]
    tag = row["tag"]

if((sid!="none") or (ind==0)):
    if(len(word_list)>0):
            list_sent.append(word_list)
            pos_tags_list = nltk.pos_tag(word_list)
            df = pd.DataFrame(pos_tags_list)
            df["id"] = sid_perm

try:
                df["tag"] = tag_list
except:
print (tag_list,word_list,len(tag_list),len(word_list))
if(sid_perm=="Sentence: 1"):
                df_all_pos = df

else:
                df_all_pos = pd.concat([df_all_pos,df],axis=0)

        word_list = []
        tag_list = []
        word_list.append(word)
        tag_list.append(tag)
else:
         word_list.append(word)
```

```
#word_list = word_list + word + " "
        tag_list.append(tag)
```

```
if(sid!="none"):
        sid_perm = sid
df_all_pos.columns = ["word","pos_tag","sid","tag"]
```

df_all is the data frame that contains the word and POS tags. Once this step is
done, you can get the lemma of the word, shape of the word, length, etc., which are not
dependent on the sentence but just the words themselves.

You also want to identify words that have either numbers or special characters in them.
You can tag these words into a separate word type before replacing the special characters
for further processing. See Listings 4-34 and 4-35.

**Listing 4-34.**

```
df_all_pos["word1"] = df_all_pos["word"]
```

```
df_all_pos["word1"] = df_all_pos.word1.str.replace('[^a-z\s]+','')
```

```
df_all_pos["word_type"] = "normal"
df_all_pos.loc[df_all_pos.word.str.contains('[0-9]+'),"word_type"] =
"number"
df_all_pos.loc[df_all_pos.word.str.contains('[^a-zA-Z\s]+'),"word_type"] =
"special_chars"
```

```
df_all_pos.loc[(df_all_pos.word_type!="normal") & (df_all_pos.word1.str.
len()==0),"word_type"] = "only_special"
```

```
deflemma_func(x):
return lemmatizer.lemmatize(x)
```

**Listing 4-35.**

```
df_all_pos["shape"] = "mixed"
df_all_pos.loc[(df_all_pos.word.str.islower()==True),"shape"]="lower"
df_all_pos.loc[(df_all_pos.word.str.islower()==False),"shape"]="upper"
```

```
df_all_pos["lemma"] = df_all_pos["word"].apply(lemma_func)
```

```
df_all_pos["length"] = df_all_pos["word"].str.len()
```

149

The code in Listing 4-36 gets the relative position of the word in the sentences and the sentence length.

### Listing 4-36.

```
df_all_pos["ind_num"] = df_all_pos.index
df_all_pos["sent_len"]=df_all_pos.groupby(["sid"])["ind_num"].transform(max)
df_all_pos1 = df_all_pos[df_all_pos.sent_len>0]
df_all_pos1["rel_position"] = df_all_pos1["ind_num"] /df_all_pos1["sent_
len"]*100
df_all_pos1["rel_position"] = df_all_pos1["rel_position"].astype('int')
```

Once you get these variables in place, you use the pandas shift functionality to get features like previous word, next word, POS tag of the previous word, POS tag of the next tag, and so on. For each feature, you call get_prev_next to get the shifted features. See Listings 4-37 and 4-38.

### Listing 4-37.

```
defget_prev_next(df_all_pos,col_imp):
    prev_col = "prev_" + col_imp
    next_col = col_imp + "_next"

    prev_col1 = "prev_prev_" + col_imp
    next_col1 = col_imp + "_next_next"

    df_all_pos[prev_col] = df_all_pos[col_imp].shift(1)
    df_all_pos.loc[df_all_pos.index==0,prev_col] = "start"

    df_all_pos[next_col] = df_all_pos[col_imp].shift(-1)
    df_all_pos.loc[df_all_pos.index==df_all_pos.sent_len,next_col] = "end"

    df_all_pos[prev_col1] = df_all_pos[col_imp].shift(2)
    df_all_pos.loc[df_all_pos.index<2,prev_col1] = "start"

    df_all_pos[next_col1] = df_all_pos[col_imp].shift(-2)
    df_all_pos.loc[(df_all_pos.sent_len-df_all_pos.index)<=1,next_col1] = "end"

return df_all_pos
```

*Listing 4-38.*

```
df_all_pos1 = get_prev_next(df_all_pos1,"word")
df_all_pos1 = get_prev_next(df_all_pos1,"lemma")
df_all_pos1 = get_prev_next(df_all_pos1,"shape")
df_all_pos1 = get_prev_next(df_all_pos1,"pos_tag")
```

You are interested in the word being an organization. So you rename all tags other than organization to no_org. Since the number of organizations in your dataset is very limited (less than 3%), you will be sampling the non-org words and keeping the org words. See Listings 4-39 and 4-40.

*Listing 4-39.*

```
t2 = df_all_pos1[df_all_pos1.tag.isna()==False]
t2["tag1"] = "no_org"
t2.loc[t2.tag.str.contains('org',case=False),"tag1"]="org"
```

*Listing 4-40.*

```
t2_neg = t2.loc[t2.tag1!="org",:]
t2_neg1 = t2_neg.sample(frac=0.1)

t2_pos = t2.loc[t2.tag1=="org",:]
t3 = pd.concat([t2_neg1,t2_pos],axis=0)
t3 = t3.reset_index()
len(t2_neg),len(t2_neg1),len(t2_pos),len(t3)
(96652, 9665, 3346, 13011)
```

Listing 4-41 splits the data into train and test sets.

*Listing 4-41.*

```
tgt = t3["tag1"]
from sklearn.model_selection import StratifiedShuffleSplit
sss = StratifiedShuffleSplit(test_size=0.2,random_state=42,n_splits=1)

for train_index, test_index in sss.split(t3, tgt):
    x_train, x_test = t3[t3.index.isin(train_index)], t3[t3.index.isin(test_index)]
    y_train, y_test = t3.loc[t3.index.isin(train_index),"tag1"], t3.loc[t3.
    index.isin(test_index),"tag1"]
```

You will also be separating the string and numeric columns. The categorical columns will get an embedding layer and the numeric columns will get added to the flattened layers of the embedding output. You will be modifying the architecture in Figure 4-11 to have numerical layers as well.

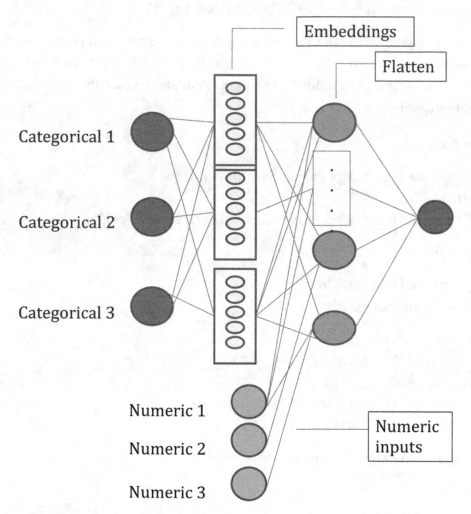

***Figure 4-11.*** *Categorical and numeric embeddings*

You connect the numeric layers to the flattened layer of the architecture. In order to do this, you keep the numeric columns separate and combine them into a matrix. See Listing 4-42.

*Listing 4-42.*

```
num_cols = ['length','sent_len', 'rel_position']

str_cols = ['word', 'pos_tag', 'shape', 'lemma', 'word_type',
'prev_word', 'word_next', 'prev_prev_word',
'word_next_next', 'prev_lemma', 'lemma_next', 'prev_prev_lemma',
'lemma_next_next', 'prev_shape', 'shape_next', 'prev_prev_shape',
'shape_next_next', 'prev_pos_tag', 'pos_tag_next', 'prev_prev_pos_tag',
'pos_tag_next_next','prev_word_type', 'word_type_next', 'prev_prev_word_type',
'word_type_next_next']

x_train_num = x_train[num_cols]
x_test_num = x_test[num_cols]
```

Now you will prepare the `train` and `test` datasets to suit the architecture at the top. Follow these steps.

1. **Tokenization and label encoding**: Inputs to the embedding layer are label-encoded categorical variables (`texts_to_sequences`). Once encoded, they also need to be padded with the required length. Since you have only single tokens per column, you keep the `max_len` to 1. The tokenization method gets the mapping of the words to corresponding indexes using the train dataset. This tokenized object is then applied to test dataset. Any unknown text in the test set will be ignored.

2. **Concatenation of all categorical input columns:** The encoded columns are concatenated into a single train and a single test matrix using a loop.

Before you get to the code, let's take a quick look at installing Keras and TensorFlow. You will be installing Keras version 2.2.4. Keras automatically uses TensorFlow as the back end. See Listings 4-43 through 4-45.

*Listing 4-43.*

```
!pip install tensorflow
!pip install keras
```

***Listing 4-44.***

```
from keras.preprocessing.text import Tokenizer
from keras.preprocessing.sequence import pad_sequences
tokenizer = Tokenizer()

defconv_str_cols(col_tr,col_te):
#print(col_tr)
    tokenizer = Tokenizer()
    tokenizer.fit_on_texts(col_tr)
    col_tr1 = tokenizer.texts_to_sequences(col_tr)
    col_te1 = tokenizer.texts_to_sequences(col_te)
    col_tr2 = pad_sequences(col_tr1, maxlen=1, dtype='int32', padding='pre')
    col_te2 = pad_sequences(col_te1, maxlen=1, dtype='int32', padding='pre')

return col_tr2,col_te2
```

***Listing 4-45.***

```
for num,i in enumerate(str_cols):

    var1,var2 = conv_str_cols(x_train[i],x_test[i])
if(num==0):

        var_all_train = var1
        var_all_test = var2

else:
        var_all_train = np.concatenate([var_all_train,var1],axis=1)
        var_all_test = np.concatenate([var_all_test,var2],axis=1)
var_all_train.shape, var_all_test.shape
((10408, 20), (2603, 20))
```

As you can see in Listing 4-45, the `train` and `test` datasets have 20 columns each, corresponding to the string columns.

- **Embeddings for categorical variables**: For each categorical variable, you compute the embeddings based on the logic suggested by fast.ai. See the Medium article at https://medium.com/@satnalikamayank12/on-learning-embeddings-for-categorical-data-using-keras-165ff2773fc9.

You're using the functional API of Keras for your purposes. It's easier to concatenate different layers using the functional API of Keras.

embedding_size = min(np.ceil((no_of_unique_cat)/2), 50 )

Listing 4-46 does the following things:

1.  Creates the embeddings for each of the categorical variables and flattens them

2.  Appends all of the independent inputs into a list (inputs)

3.  Appends the flattened outputs into another list called a dense list object (outputs)

4.  Creates separate numeric input and appends it to the dense list object

5.  Connects the numeric input to `num_inp1` to the dense list object

6.  Concatenates all the embeddings and the numeric input to one layer

7.  `df_list` is the list of all categorical and numeric train inputs.

8.  `df_list_test` is the list of all categorical and numeric test inputs.

## *Listing 4-46.*

```
import keras
from keras.models import Sequential
from keras.layers import Dense,Flatten,Concatenate
from keras.layers.merge import concatenate
from keras.layers import Input, Dense, Dropout, Flatten
from keras.models import Model
from keras.utils import to_categorical
from keras.optimizers import Adam,SGD
from keras.layers import Embedding
from keras.layers import Reshape
from keras.layers import add
```

```python
embed_size=0
models = []
inputs = []
outputs = []
dense = []
df_list = []
df_list_test=[]
for num,categoical_var in enumerate(range(var_all_train.shape[1]) ):

    model = Sequential()
    no_of_unique_cat =np.max(var_all_train[:,num]) + 1

    embedding_size = min(np.ceil((no_of_unique_cat)/2), 50 )
    embedding_size = int(embedding_size)
    embed_size = embed_size + embedding_size
    vocab  = no_of_unique_cat+1

    A1 = Input(shape=(1,))
    A2 = Embedding(vocab,embedding_size,input_length = 1)(A1)

    A3 = Flatten()(A2)
    dense.append(A3)
    inputs.append(A1)

    df_list.append(var_all_train[:,num])
    df_list_test.append(var_all_test[:,num])

num_shape = x_train_num.shape[1]
embed_size = embed_size + num_shape
num_inp = Input(shape=(num_shape,))
num_inp1 = Dense(num_shape,activation = 'relu')(num_inp)
dense.append(num_inp1)
inputs.append(num_inp)

st_size = int(embed_size/2)

df_list.append(x_train_num)
df_list_test.append(x_test_num)

merge_one = concatenate(dense)
```

The layer merge_one will now be connected to other dense layers. Now, you reduce the number of nodes in each of the following layers by two and obtain a list of layer parameters. layers_list contains the architecture of the neural network after the merge_one. The function get_nn_mod lays out the rest of the architecture from merge_one based on the layers_list. We start with st_size1 initialized in Listing 4-46. This is the size of the final concatenated layer divided by 2. Model is a function that finally takes the initial inputs and the final outputs. The model object is now ready to be compiled. See Listings 4-47 and 4-48.

***Listing 4-47.***

```
####flat layers list
layers_list = []
st_size1 = st_size
while (st_size1>10):
    st_size1 = int(st_size1/2)
    layers_list.append(st_size1)

Layers_list
[148, 74, 37, 18, 9]
```

***Listing 4-48.***

```
defget_nn_mod(list_layers,input1,dp,inputs):
    layers = []

for num,i in enumerate(list_layers):
        print(num,i)

if(num==0):
            input_orig = input1

else:
            input1 = Dense(i, activation-'relu')(input1)
            input1 = Dropout(dp)(input1)

    input_last = Dense(2, activation='softmax')(input1)

    model = Model(inputs=inputs, outputs=input_last)

    opt = SGD(lr=0.01, clipnorm=1.)
```

```
# Compile model
    model.compile(optimizer=opt, loss='categorical_crossentropy',
                  metrics=['accuracy'])
return model
```

You can now build the model and check the summary and the accuracies. See Listing 4-49.

**Listing 4-49.**

```
final_model = get_nn_mod(layers_list,merge_one,0.6,inputs)
0 148
1 74
2 37
3 18
4 9
```

Now convert the dependent varibable to categorical with the code in Listings 4-50 and 4-51. See Figure 4-12.

**Listing 4-50.**

```
from keras.utils import to_categorical
from sklearn.preprocessing import LabelEncoder
le = LabelEncoder()
y_train1 = le.fit_transform(y_train)
y_test1 = le.fit_transform(y_test)

y_train2 = to_categorical(y_train1)
y_test2 = to_categorical(y_test1)
```

**Listing 4-51.**

```
pd.Series(y_train1).value_counts()
0    7731
1    2677
dtype: int64

final_model.fit(df_list,y_train2, batch_size=10, epochs=30,
        verbose=2)
```

```
pred1=final_model.predict(df_list_test)
pred = pred1.argmax(axis=-1)
from sklearn.metrics import accuracy_score
from sklearn.metrics import f1_score
ac1 = accuracy_score(y_test1, pred)
print (ac1, f1_score(y_test1, pred, average='macro'))
0.9193238570879754 0.896944708384236

from sklearn.metrics import confusion_matrix
cmat = pd.DataFrame(confusion_matrix(y_test1, pred, labels=[0,1], sample_
weight=None))
cmat.columns = rows_name
cmat["act"] = rows_name
cmat
```

|   | 0 | 1 | act |
|---|---|---|---|
| 0 | 1803 | 131 | 0 |
| 1 | 79 | 590 | 1 |

*Figure 4-12.*

# Applying the Model

So you now have a decent model with a good accuracy. You can now apply the model to the NER news data set (Data mention in Listing 4-1). See Listing 4-52.

*Listing 4-52.*

```
import pandas as pd
import nltk
from nltk.stem import WordNetLemmatizer
pd.options.display.max_colwidth = 1000
```

Step 1: After reading the initial NER dataset, you split the sentences into words and the corresponding POS tags like you did for the training set. See Listing 4-53.

***Listing 4-53.***

```
lemmatizer = WordNetLemmatizer()
for num,i in enumerate(t2.loc[:,"imp_col"]):

    for num1,j in enumerate(nltk.sent_tokenize(i)):
        pos_tags_list = nltk.pos_tag(nltk.word_tokenize(j))
        df = pd.DataFrame(pos_tags_list)
        df.columns = ["word","pos_tag"]
        df["sid"] = num1
        df["sid1"] = num
if(num==0):
            df_all = df
else:
            df_all = pd.concat([df_all,df],axis=0)
df_all_pos = df_all
```

Step 2: Create word type variables, a lemma of words, and the relative length variables. See Listings 4-54 and 4-55.

***Listing 4-54.***

```
df_all_pos["word1"] = df_all_pos["word"]

df_all_pos["word1"] = df_all_pos.word1.str.replace('[^a-z\s]+','')

df_all_pos["word_type"] = "normal"
df_all_pos.loc[df_all_pos.word.str.contains('[0-9]+'),"word_type"] = "number"
df_all_pos.loc[df_all_pos.word.str.contains('[^a-zA-Z\s]+'),"word_type"] =
"special_chars"

df_all_pos.loc[(df_all_pos.word_type!="normal") & (df_all_pos.word1.str.
len()==0),"word_type"] = "only_special"
```

***Listing 4-55.***

```
df_all_pos["shape"] = "mixed"
df_all_pos.loc[((df_all_pos.word.str.islower()==True) & (df_all_pos.word_
type=="normal")),"shape"]="lower"
```

```
df_all_pos.loc[((df_all_pos.word.str.islower()==False) & (df_all_pos.word_ty
pe=="normal")),"shape"]="upper"

deflemma_func(x):
return lemmatizer.lemmatize(x)

df_all_pos["lemma"] = df_all_pos["word"].apply(lemma_func)

df_all_pos["length"] = df_all_pos["word"].str.len()

df_all_pos["ind_num"] = df_all_pos.index
df_all_pos["sent_len"]=df_all_pos.groupby(["sid","sid1"])["ind_num"].
transform(max)
df_all_pos1 = df_all_pos[df_all_pos.sent_len>0]
df_all_pos1["rel_position"] = df_all_pos1["ind_num"] /df_all_pos1["sent_
len"]*100
df_all_pos1["rel_position"] = df_all_pos1["rel_position"].astype('int')
```

Step 3: Now get the positional variables like previous, next, etc. See Listings 4-56 and 4-57.

***Listing 4-56.***

```
def get_prev_next(df_all_pos,col_imp):
    prev_col - "prev_" + col_imp
    next_col = col_imp + "_next"

    prev_col1 = "prev_prev_" + col_imp
    next_col1 = col_imp + "_next_next"

    df_all_pos[prev_col] = df_all_pos[col_imp].shift(1)
    df_all_pos.loc[df_all_pos.index==0,prev_col] = "start"

    df_all_pos[next_col] = df_all_pos[col_imp].shift(-1)
    df_all_pos.loc[df_all_pos.index==df_all_pos.sent_len,next_col] = "end"

    df_all_pos[prev_col1] = df_all_pos[col_imp].shift(2)
    df_all_pos.loc[df_all_pos.index<2,prev_col1] = "start"

    df_all_pos[next_col1] = df_all_pos[col_imp].shift(-2)
    df_all_pos.loc[(df_all_pos.sent_len-df_all_pos.index)<=1,next_col1] = "end"

return df_all_pos
```

*Listing 4-57.*

```
df_all_pos1 = get_prev_next(df_all_pos1,"word")
df_all_pos1 = get_prev_next(df_all_pos1,"lemma")
df_all_pos1 = get_prev_next(df_all_pos1,"shape")
df_all_pos1 = get_prev_next(df_all_pos1,"pos_tag")
df_all_pos1 = get_prev_next(df_all_pos1,"word_type")
```

Step 4: Categorize string and number columns. See Listing 4-58.

*Listing 4-58.*

```
num_cols = ['length','sent_len', 'rel_position']

str_cols = ['word', 'pos_tag', 'shape', 'lemma', 'word_type',
'prev_word', 'word_next', 'prev_prev_word',
'word_next_next', 'prev_lemma', 'lemma_next', 'prev_prev_lemma',
'lemma_next_next', 'prev_shape', 'shape_next', 'prev_prev_shape',
'shape_next_next', 'prev_pos_tag', 'pos_tag_next', 'prev_prev_pos_tag',
'pos_tag_next_next','prev_word_type', 'word_type_next', 'prev_prev_word_type',
'word_type_next_next']

x_test_num = df_all_pos1[num_cols]
x_test = df_all_pos1
```

Step 5: Load the model files and tokenization files. The x_test dataset is suitably modified to input for the model run. See Listing 4-59.

*Listing 4-59.*

```
import pickle
defpick_in(obj_name):
    fl_out1 = fl_out + "/" + obj_name
    pickle_in = open(fl_out1,"rb")
    mod1= pickle.load(pickle_in)

return mod1

fl_out =  "model_punc"
pikl_list = ["tkn_list"]
for i in pikl_list:
    globals()[i] = pick_in(i)
```

```
from keras.models import load_model
final_model = load_model('model_punc/model_ner.h5')
```

Now convert the string dataset to tokenized inputs for embedding layers. Next, concatenate the final string array with all the numeric data into a single dataset. See Listings 4-60 and 4-61.

*Listing 4-60.*

```
from keras.preprocessing.text import Tokenizer
from keras.preprocessing.sequence import pad_sequences
defconv_str_cols(col_te,num):

    col_te1 = tkn_list[num].texts_to_sequences(col_te)

    col_te2 = pad_sequences(col_te1, maxlen=1, dtype='int32', padding='pre')

return col_te2

import numpy as np
for num,i in enumerate(str_cols):

    var1 = conv_str_cols(x_test[i],num)

if(num==0):
        var_all_test = var1
else:
        var_all_test = np.concatenate([var_all_test,var1],axis=1)
```

*Listing 4-61.*

```
df_list_test=[]
for num,categoical_var in enumerate(range(var_all_test.shape[1]) ):
    df_list_test.append(var_all_test[:,num])
df_list_test.append(x_test_num)
```

Step 6: Apply the model and identify the predicted organization from the news corpus. See Listing 4-62 and Figure 4-13.

***Listing 4-62.***

```
pred1=final_model.predict(df_list_test)
pred = pred1.argmax(axis=-1)
pd.Series(pred).value_counts()
```

```
                    0      1941596
                    1       135599
                    dtype: int64
```

***Figure 4-13.*** *Distribution of word predicted as "organization" and "organization" entities*

Since you trained the model on a biased set of organizations, you do expect some false positives in your dataset. Now you collect the identified organization words and validate them with a list of possibly non-organization names. In Listing 4-63, you concatenate consecutive organization words in the same sentence to form one full word. Since your model is at word level, you are concatenating to form a chunk. See Figure 4-14.

***Listing 4-63.***

```
x_test["pred"] = pred

list_set_words = []
sid_prev = 0
fg=0
sid1_list = []
for num,row in x_test[x_test.pred==1].iterrows():
    wrd = row["word"]
    ind_num = row["ind_num"]
    sid = row["sid"]
    sid1 = row["sid1"]
if((num==0) or (sid_prev!=sid)):
        wrd_prev = wrd

else:
if((ind_num-ind_num_prev)==1):
            wrd_prev = wrd_prev + " " + wrd
            fg=1
```

```
else:
if(fg==1):
                list_set_words.append(wrd_prev)
                sid1_list.append(sid1_prev)
                fg=0
            wrd_prev = wrd
    ind_num_prev = ind_num
    sid_prev = sid
    sid1_prev = sid1

df_all_ents = pd.DataFrame(sid1_list)
df_all_ents.columns = ["sid1"]
df_all_ents["ner_words"] = list_set_words

df_all_ents.head()
```

| | sid1 | ner_words |
|---|---|---|
| 0 | 1 | Malaika Arora Khan |
| 1 | 2 | Indira Gandhi Institute of Medical |
| 2 | 2 | Dr Manish Mandal |
| 3 | 2 | Dr |
| 4 | 2 | All India Institute of Medical Sciences |

*Figure 4-14.*

As expected, the dataset with all collected named entities contains named entities with non-organization words. Also note that the same document (sid1) can have multiple named entities. However, when you started, you had a candidate set of 2MN words in the corpus and this has come down to 135K possible words for organizations. The dataset further reduced to 22K phrases in df_all_ents. You write this to a file called only_ner_tagged.csv. You can now run this through a known negative corpus to eliminate non-organization words. This is detailed in the following step.

**Step 7: Getting a negative corpus:** Given that you are dealing with a news corpus and you may see a mix of organization and non-organization words, you can use a public dataset to remove celebrities, politicians, or names of places like Chennai, Mumbai, etc. You can use Wikidata for the same. Wikidata (`www.wikidata.org/wiki/Wikidata:Main_Page`) is an open source project that stores underlying information from wiki projects in a RDF format. The Resource Description Framework is the data model of the semantic web and you can use SPARQL to query data stored in RDF. You use SPARQL on the Wikidata corpus to get a list of non-organization entities.

So use SPARQL to get the following lists from Wikidata:

- **Celebrities of India**: Famous personalities of India are retrieved through the query in Listing 4-64.

### Listing 4-64.

```
SELECT ?item ?itemLabel WHERE { ?item wdt:P31 wd:Q5; wdt:P27 wd:Q668;
SERVICE wikibase:label { bd:serviceParam wikibase:language "en". } }
```

- **Places in India**: Cities, states, towns, and countries of the world (Listing 4-65).

### Listing 4-65.

```
SELECT ?city ?cityLabel ?state WHERE {

  {?city (wdt:P31/(wdt:P279*)) wd:Q3957}
  UNION
  {?city (wdt:P31/(wdt:P279*)) wd:Q515}
  UNION
  {?city (wdt:P31/(wdt:P279*)) wd:Q149621}
  UNION
  {?city (wdt:P31/(wdt:P279*)) wd:Q13390680 }
    ?city wdt:P17 wd:Q668
SERVICE wikibase:label { bd:serviceParam wikibase:language "[AUTO_LANGUAGE],en". }
}
```

- Political positions like prime minister, chief minister, etc. (Listing 4-66)

*Listing 4-66.*

```
SELECT ?postion ?positionLabel   WHERE {

  {?position (wdt:P31/(wdt:P279*)) wd:Q4164871;
          wdt:P1001 ?juris.
          ?juris wdt:P31 wd:Q13390680.}

  UNION

  {
    ?position wdt:P31 wd:Q294414.
    ?position wdt:P17 wd:Q668.}

  UNION

  {
     ?position wdt:P31 wd:Q4164871.
    ?position wdt:P17 wd:Q668.
    }

  SERVICE wikibase:label { bd:serviceParam wikibase:language "[AUTO_
  LANGUAGE],en". }
}
```

Step 8: After obtaining the lists from step 7, you now proceed to validate the 22K dataset in step 6. You compare the set of files you have and eliminate common words using a cosine similarity. The detailed process is described in the following steps.

Step 8.1: Read your only_ner_tagged.csv in another notebook. Process the data into two lists, one with spaces and one without spaces. See Listing 4-67.

*Listing 4-67.*

```
import pandas as pd
t1 = pd.read_csv("only_ner_tagged.csv")
t2 = t1[t1.ner_words.str.len()>=4]
t2["ner_words1"] = t2["ner_words"].str.lower()
t2["ner_words2"] = t2["ner_words1"].str.replace(" ","")
ner_list_space = t2["ner_words1"].unique()
ner_list = pd.Series(ner_list_space).str.replace(" ","")
```

```
len(t1),len(t2),len(ner_list_space),len(ner_list)
(22762, 21726, 12239, 12239)
```

Step 8.2: Read the files generated in step 7. We also print out the files from wikidata so that the readers get an idea of the data. See Listings 4-68 through 4-71 and Figures 4-14 through 4-17.

*Listing 4-68.*

```
wiki_names = pd.read_csv("wikidata\persons_india.csv")
wiki_places = pd.read_csv("wikidata\city_state_country.csv")
wiki_positions = pd.read_csv("wikidata\positions_india.csv")
```

*Listing 4-69.*

```
#celebrity names from wikidata
wiki_names.head()
```

|   | url | itemLabel |
|---|-----|-----------|
| 0 | http://www.wikidata.org/entity/Q55713 | Anand Yadav |
| 1 | http://www.wikidata.org/entity/Q55716 | Bhargavaram Viththal Varerkar |
| 2 | http://www.wikidata.org/entity/Q55719 | Sarojini Vaidya |
| 3 | http://www.wikidata.org/entity/Q55735 | Aroon Tikekar |
| 4 | http://www.wikidata.org/entity/Q55744 | Vijay Tendulkar |

*Figure 4-15.*

*Listing 4-70.*

```
##places in india and countries of word
wiki_places.head()
```

| | url | itemLabel |
|---|---|---|
| 0 | http://www.wikidata.org/entity/Q1154 | Jamnagar |
| 1 | http://www.wikidata.org/entity/Q9461 | Khedbrahma |
| 2 | http://www.wikidata.org/entity/Q9894 | Dharmapuri |
| 3 | http://www.wikidata.org/entity/Q11854 | Rajkot |
| 4 | http://www.wikidata.org/entity/Q13221 | Tezu |

*Figure 4-16.*

*Listing 4-71.*

```
##positions of public importance in india
wiki_positions.head()
```

| | url | itemLabel1 | itemLabel |
|---|---|---|---|
| 0 | NaN | Chief Minister of Telangana | Chief Minister |
| 15 | NaN | of Maharashtra | |
| 29 | NaN | governor of Telangana | governor |
| 32 | NaN | Member of Telangana Legislative Council | Member |
| 33 | NaN | Governor of Punjab | Governor |

*Figure 4-17.*

Step 8.3: Process the columns and combine the files into a single dataframe. You then create non-duplicate lists with and without spaces. See Listings 4-72 and 4-73.

*Listing 4-72.*

```
wiki_names.columns = ["url","itemLabel"]
wiki_places.columns = ["url","itemLabel"]
wiki_positions.columns = ["url","itemLabel1"]
```

```
##processing words like "of Gujarat","of Delhi" to arrive at better similarity
wiki_positions["itemLabel"] = wiki_positions["itemLabel1"].str.replace('of
[a-zA-Z\s]+','')

wiki_names = wiki_names.drop_duplicates(["itemLabel"])
wiki_places = wiki_places.drop_duplicates(["itemLabel"])
wiki_positions = wiki_positions.drop_duplicates(["itemLabel"])

wiki_all = pd.concat([wiki_names,wiki_places,wiki_positions],axis=0)
```

***Listing 4-73.***

```
wiki_all["itemLabel1"] = wiki_all["itemLabel"].str.lower()
item_list_space = pd.Series(wiki_all["itemLabel1"].unique())
wiki_all["itemLabel2"] = wiki_all["itemLabel1"].str.replace(" ","")

item_list = pd.Series(item_list_space).str.replace(" ","")
```

Step 8.4: Now vectorize the lists in steps 8.1 and 8.3. The lists with spaces get vectorized using a word vectorizer (ner_list, item_list). The lists without spaces gets vectorized as a character vectorizer (ner_list_space, item_list_space). You then concatenate the outputs into two single arrays: one for your shortlisted named entities and the other for the Wikidata corpus. See Listings 4-74 and 4-75.

***Listing 4-74.***

```
from sklearn.feature_extraction.text import TfidfVectorizer

vectorizer = TfidfVectorizer(analyzer = 'char',ngram_range = (4,6),min_
df=0.0001)
ner_vect_char = vectorizer.fit_transform(ner_list)
wiki_vect_char = vectorizer.transform(item_list)

vectorizer1 = TfidfVectorizer(analyzer = 'word',ngram_range = (1,1),min_
df=0.0001)
ner_vect_word = vectorizer1.fit_transform(ner_list_space)
wiki_vect_word = vectorizer1.transform(item_list_space)
```

*Listing 4-75.*

```
from scipy.sparse import  hstack
ner_vect = hstack([ner_vect_char,ner_vect_word])
wiki_vect = hstack([wiki_vect_char,wiki_vect_word])
```

Step 8.5: Now for the cosine similarity of ner_vect and wiki_vect. See Listing 4-76.

*Listing 4-76.*

```
from sklearn.metrics.pairwise import cosine_similarity
sim_vec = cosine_similarity(ner_vect,wiki_vect,dense_output=False)
```

Step 8.6: You consider it a match if the cosine similarity is greater than 0.6. Since you have a sparse matrix, you convert that to a coordinate format (coo). Coo formats store data in three arrays: one for row, column, and data. See Listing 4-77 and Figure 4-18.

*Listing 4-77.*

```
condition = (sim_vec >=.6)
coo = condition.tocoo()

df_matches = pd.DataFrame(columns = ["wiki_match_label","ner_match_label"])
wiki_match_label = [item_list[i] for i in coo.col]

ner_match_label = [ner_list[i] for i in coo.row]

df_matches["wiki_match_label"] = wiki_match_label
df_matches["ner_match_label"] = ner_match_label

df_matches.head()
```

| | wiki_match_label | ner_match_label |
|---|---|---|
| 0 | malaikaarora | malaikaarorakhan |
| 1 | manishmanikpuri | drmanishmandal |
| 2 | kashmirsingh | kashmir |
| 3 | rupsa,india | india |
| 4 | bassi,india | india |

*Figure 4-18.*

171

Step 8.4: Match with the original dataset and get the non-matched entities as your final filtered organization entities. See Listing 4-78.

### Listing 4-78.

```
t3 = pd.merge(t2,df_matches, left_on = "ner_words2", right_on = "ner_match_
label", how = "left")
len(t3[t3.ner_match_label.isna()==True])
13149
```

Your dataset has been reduced by half and you can see a quick distribution of the top words in your list. See Listing 4-79 and Figure 4-19.

### Listing 4-79.

```
t3.loc[t3.ner_match_label.isna()==True,"ner_words1"].value_counts()
```

```
:  supreme court                         222
   congress                              141
   air india                              94
   world cup                              86
   open sans                              77
   aam aadmi party                        56
   india today                            54
   police                                 53
   bharatiya janata party                 52
   also                                   51
   party                                  50
   indian army                            40
   army                                   32
   us president donald trump              29
   income tax                             29
   samajwadi party                        29
   high court                             27
   bombay high court                      27
   april                                  26
   rajya sabha                            26
   university                             25
   lok sabha                              25
   state bank                             24
   president donald trump                 24
```

### Figure 4-19.

You can see a lot of organizations in this list. There are 115 unique words if you apply a filter of count >= 10. Listing 4-81 is a script that identifies the percentage of English words in every row here. Typical organization names will have less English words and more of "names" like words. You are using a library named enchant `https://pypi.org/project/pyenchant/` for this purpose This will further help you deep dive in the next step. Also, see Listing 4-81.

### Listing 4-80.

```
shrt_words = pd.DataFrame(t3.loc[t3.ner_match_label.isna()==True,"ner_
words1"].value_counts()).reset_index()
shrt_words.columns = ["words","count"]
```

### Listing 4-81.

```
word_checks = enchant.Dict("en_US")
defchk_eng(word):
    tot_flag = 0
for i in word.split():
        flag = word_checks.check(i)
if(flag==True):
            tot_flag - tot_flag +1

return tot_flag

shrt_words["eng_ind"] = shrt_words["words"].apply(chk_eng)
shrt_words["word_cnt"] = (shrt_words["words"].str.count(" "))+1
shrt_words["percent"] = shrt_words["eng_ind"]/shrt_words["word_cnt"]
shrt_words.to_clipboard()
```

Step 8.5: Getting back to where you started, the bank is interested in monitoring organizations that are mentioned in public forums for credit tracking or fraud monitoring purposes. You can now look at these 115 words (top with count >= 10) or all the words (value counts of the t3 dataset) or shrt_words (removing fully English words) and possibly manually skim the list and identify the organizations of interest. Note: Since you are dealing with news articles in all sections, it is very common to get all mentions of public organizations like political parties or public organizations (Example, "Indian Army"). You will have to skim through the list and identify private organizations of interest or you can match against a lookup list here.

By manually going through the lists generated in step 8.4, I came down to the organizations listed in Table 4-5.

*Table 4-5.*

| |
|---|
| **air india** |
| **gandhi institute of medical** |
| **red bull** |
| **united airlines** |
| **qatar airways** |
| **patanjali research institute** |
| **jet airways** |
| **millennium school** |
| **facebook** |
| microsoft |
| cbfc |
| maruti celerio |
| facebook india md |
| sony xperia xz premium |
| foxx |
| uber ceo travis kalanick |
| bharti airtel |
| airtel |
| change.org |
| uber india president amit jain |
| **emami** |

Step 9: You need to identify the original articles with the shortlisted organizations in step 8.5. See Listing 4-82.

***Listing 4-82.***

```
###modified uber indi president to uber india
df_all_short_list = pd.read_csv("interest_words_tag.csv",header=None)
df_all_short_list.columns = ["words_interest"]
list_short = list(df_all_short_list["words_interest"].unique())
```

You match the shortlisted words with the ones in the corpus in the column `imp_col` and you take the matches. You have finally arrived at 216 articles from your initial corpus of 4396 sentences. See Listings 4-83 and 4-84.

### Listing 4-83.

```
list_interest_all = []
for i in t2["imp_col"].str.lower():
    wrd_add = ''
for j in list_short:
if(i.find(j)>=0):
            wrd_add = j
break;
    list_interest_all.append(wrd_add)
```

### Listing 4-84.

```
t2["short_list"] = list_interest_all
t3 = t2[t2.short_list.str.len()>=3]
len(t3)
216
```

Step 10: You now get the sentiment of the sentences. For a quick illustration, you will use the Vader Sentiment package but this can be done using any sentiment technique in Chapter 3. See Listings 4-85 and 4-86 and Figure 4-20.

### Listing 4-85.

```
from vaderSentiment.vaderSentiment import SentimentIntensityAnalyzer
analyser = SentimentIntensityAnalyzer()

defget_vader_score(sents):
    score = analyser.polarity_scores(sents)
return score['compound']

t3["score"] = t3["imp_col"].apply(get_vader_score)
```

*Listing 4-86.*

```
sent_agg = t3.groupby(['short_list'], as_index = False)['score'].mean()
sent_agg
```

| | short_list | score |
|---|---|---|
| 0 | air india | -0.123904 |
| 1 | airtel | 0.964140 |
| 2 | bharti airtel | 0.284985 |
| 3 | cbfc | -0.197989 |
| 4 | change.org | -0.776450 |
| 5 | emami | 0.959300 |
| 6 | facebook | 0.000423 |
| 7 | foxx | 0.029850 |
| 8 | gandhi institute of medical | -0.092130 |
| 9 | jet airways | -0.851844 |
| 10 | maruti celerio | -0.998500 |
| 11 | microsoft | 0.406889 |
| 12 | millennium school | -0.941900 |
| 13 | patanjali research institute | 0.799800 |
| 14 | qatar airways | -0.338900 |
| 15 | red bull | 0.917275 |
| 16 | sony xperia xz premium | 0.969650 |
| 17 | uber india | 0.926200 |

*Figure 4-20.*

# Other Use Cases of NLP in Banking, Financial Services, and Insurance

The banking, financial services, and insurance industries collect a lot of structured data from customers. For large banks, the data collected per user can run into hundreds of thousands of columns. Hence, even if NLP is used, it is used as a few of the variables in the model. The key sources of NLP data could include customer care (voice and text), sales representative notes, relationship manager notes, customer transaction notes, survey feedback, and recent external data (social media and news corpus data). In the case of insurance, claims notes get processed. Using this data, NLP is used to identify fraud, risk, attrition, targeting for marketing campaigns, sales malpractices, claims process, recommendation of investment, portfolio management, etc.

## SMS Data

Another source of text data that has become widely used with financial institutions is user SMS data. There are personal financial management companies that mine user SMS data and get the state of a user's need and financial health. A user's SMS data (to be used with user consent) offers rich information of that user's shopping and financial preferences.

## Natural Language Generation in Banks

Automated insight generation is an upcoming field that has seen adoption by investment companies and financial organizations that do a lot of report crunching. Given a bunch of graphs and charts, we can generate a meaningful automated narrative out of them using some of the natural language generation techniques. Virtual assistants are now occupying center stage in financial institutions. Using historic customer support data, we can develop conversation agents that can understand and respond appropriately to customers.

177

Table 4-6 lists the packages used in this chapter.

**Table 4-6.**  *Packages Used in This Chapter*

| pack_name | Version |
| --- | --- |
| absl-py | 0.7.1 |
| astor | 0.8.0 |
| attrs | 19.1.0 |
| backcall | 0.1.0 |
| bleach | 3.1.0 |
| blis | 0.2.4 |
| boto3 | 1.9.199 |
| boto | 2.49.0 |
| botocore | 1.12.199 |
| certifi | 2019.3.9 |
| chardet | 3.0.4 |
| click | 7.1.2 |
| colorama | 0.4.1 |
| cycler | 0.10.0 |
| cymem | 2.0.2 |
| dataclasses | 0.7 |
| decorator | 4.4.0 |
| defusedxml | 0.6.0 |
| docopt | 0.6.2 |
| docutils | 0.14 |
| eli5 | 0.9.0 |
| en-core-web-md | 2.1.0 |
| en-core-web-sm | 2.1.0 |

*(continued)*

***Table 4-6.*** (*continued*)

| pack_name | Version |
|---|---|
| entrypoints | 0.3 |
| fake-useragent | 0.1.11 |
| filelock | 3.0.12 |
| fuzzywuzzy | 0.18.0 |
| gast | 0.2.2 |
| gensim | 3.6.0 |
| gensim | 3.8.0 |
| graphviz | 0.11.1 |
| grpcio | 1.21.1 |
| h5py | 2.9.0 |
| idna | 2.8 |
| imageio | 2.9.0 |
| inflect | 2.1.0 |
| ipykernel | 5.1.1 |
| ipython-genutils | 0.2.0 |
| ipython | 7.5.0 |
| jedi | 0.13.3 |
| jinja2 | 2.10.1 |
| jmespath | 0.9.4 |
| joblib | 0.13.2 |
| jsonschema | 3.0.1 |
| jupyter-client | 5.2.4 |
| jupyter-core | 4.4.0 |
| keras-applications | 1.0.8 |
| keras-preprocessing | 1.1.0 |

(*continued*)

***Table 4-6.*** (*continued*)

| pack_name | Version |
|---|---|
| keras | 2.2.4 |
| kiwisolver | 1.1.0 |
| lime | 0.2.0.1 |
| markdown | 3.1.1 |
| markupsafe | 1.1.1 |
| matplotlib | 3.1.1 |
| mistune | 0.8.4 |
| mlxtend | 0.17.0 |
| mock | 3.0.5 |
| murmurhash | 1.0.2 |
| nbconvert | 5.6.0 |
| nbformat | 4.4.0 |
| networkx | 2.4 |
| nltk | 3.4.3 |
| num2words | 0.5.10 |
| numpy | 1.16.4 |
| oauthlib | 3.1.0 |
| packaging | 20.4 |
| pandas | 0.24.2 |
| pandocfilters | 1.4.2 |
| parso | 0.4.0 |
| pickleshare | 0.7.5 |
| pillow | 6.2.0 |
| pip | 19.1.1 |
| plac | 0.9.6 |

(*continued*)

***Table 4-6.*** (*continued*)

| pack_name | Version |
|---|---|
| preshed | 2.0.1 |
| prompt-toolkit | 2.0.9 |
| protobuf | 3.8.0 |
| pybind11 | 2.4.3 |
| pydot | 1.4.1 |
| pyenchant | 3.0.1 |
| pygments | 2.4.2 |
| pyparsing | 2.4.0 |
| pyreadline | 2.1 |
| pyrsistent | 0.15.2 |
| pysocks | 1.7.0 |
| python-dateutil | 2.8.0 |
| pytrends | 1.1.3 |
| pytz | 2019.1 |
| pywavelets | 1.1.1 |
| pyyaml | 5.1.1 |
| pyzmq | 18.0.0 |
| regex | 2020.6.8 |
| requests-oauthlib | 1.2.0 |
| requests | 2.22.0 |
| s3transfer | 0.2.1 |
| sacremoses | 0.0.43 |
| scikit-image | 0.17.2 |
| scikit-learn | 0.21.2 |

(*continued*)

***Table 4-6.*** (*continued*)

| pack_name | Version |
|---|---|
| scipy | 1.3.0 |
| sentencepiece | 0.1.92 |
| setuptools | 41.0.1 |
| shap | 0.35.0 |
| six | 1.12.0 |
| sklearn | 0 |
| smart-open | 1.8.4 |
| spacy | 2.1.4 |
| srsly | 0.0.6 |
| stop-words | 2018.7.23 |
| tabulate | 0.8.3 |
| tensorboard | 1.13.1 |
| tensorflow-estimator | 1.13.0 |
| tensorflow | 1.13.1 |
| termcolor | 1.1.0 |
| testpath | 0.4.2 |
| textblob | 0.15.3 |
| thinc | 7.0.4 |
| tifffile | 2020.7.22 |
| tokenizers | 0.7.0 |
| tornado | 6.0.2 |
| tqdm | 4.32.1 |
| traitlets | 4.3.2 |
| transformers | 2.11.0 |

(*continued*)

***Table 4-6.*** (*continued*)

| pack_name | Version |
|---|---|
| tweepy | 3.8.0 |
| typing | 3.7.4 |
| urllib3 | 1.25.3 |
| vadersentiment | 3.2.1 |
| wasabi | 0.2.2 |
| wcwidth | 0.1.7 |
| webencodings | 0.5.1 |
| werkzeug | 0.15.4 |
| wheel | 0.33.4 |
| wincertstore | 0.2 |
| word2number | 1.1 |
| wordcloud | 1.5.0 |
| xgboost | 0.9 |

# CHAPTER 5

# NLP in Virtual Assistants

Virtual assistants are intelligent systems that respond to user queries. They can also hold conversations with users. We can trace the origin of virtual assistants to the 1950s when the Turing test was used to distinguish human conversations from machine conversations. NLP has undergone remarkable changes since then, especially in the last decade or so. Natural Language generation techniques where in a machine synthesis responses are prevalent in many organizations today. The chapters so far in this book have been dealing with NLU, or natural language understanding. You saw various ways of extracting intents, sentiments, etc. In NLG, machine learning algorithms are used to generate sentences. You will explore an example of natural language generation for a bot in this chapter.

## Types of Bots

Bots are classified into two main categories: general purpose and goal-oriented. General purpose bots are meant for fun or non-specific goals. Siri is a general purpose bot. You can ask simple questions like "How are you?" and "What's the weather?" and you can watch the bot give fun-filled answers. Here, the users do not have a goal in mind except to strike up a conversation. An example from the Curie bot which is a general purpose bot (`https://intentico.ai/`) is shown in Figure 5-1.

© Mathangi Sri 2021
M. Sri, *Practical Natural Language Processing with Python*, https://doi.org/10.1007/978-1-4842-6246-7_5

***Figure 5-1.*** *Talking to Curie*

Goal-oriented bots tackle very specific questions and are expected to provide very precise answers like checking flight status or getting options for hotel booking. An example is given in Table 5-1.

***Table 5-1.*** *A Goal-Oriented Bot*

| | |
|---|---|
| **Bot** | How can I help you? |
| **Customer** | Want to book a table for 2 |
| **Bot** | Sure. Which location are you looking at? |
| **Customer** | I stay in east of avenue street and hence looking for places close by |

(*continued*)

***Table 5-1.*** (*continued*)

| Bot | Which cuisine would you prefer? |
|---|---|
| Customer | Fine Dine Indian Restaurant |
| Bot | Great to know that. I can book table for two at Punjabi Frill at 8PM this evening. |
| Customer | Perfect |
| Bot | Have a great day. It's a pleasure to serve you. |

The answers by the bot have to be precise and should solve the customer's problem. Any deviation from this behavior by the bot would lead to a very bad user experience.

We would be seeing in this chapter, 2 methods to train bots. One is a very classic approach and is more customized for business use cases and the other one is more focused on generating a conversational corpus. At the end of this chapter, you will use Bidirectional Encoder Representations from Transformers (BERT) language models from Google for the classic approach. A list of packages used is provided in Table 5-6 at the end of the chapter.

# The Classic Approach

Consider an organization that is already running chat- or voice-based customer service. (We generally use this scenario as a starting point.) We divide this corpus into intents. For each intent, we identify the workflow that needs to be followed by the bot: the next set of questions, integration with the IT systems, or any other steps to handle the intent well. You will see an example herewith.

I have created a sample dataset of questions that people commonly ask customer service at a credit card company. The intent here identifies which workflow the bot should follow. For demo purposes, I have come up with the next question following the intent. You will build a simple classifier using long short-term memory architecture (LSTM) followed by a regular neural network. Before you delve into LSTM, let's have a quick look at the data. See Listing 5-1 and Figure 5-2.

***Listing 5-1.*** The Data

```
import pandas as pd
t1 = pd.read_csv("bank4.csv",encoding = 'latin1')
t1.head()
```

|   | Line | Final intent |
|---|------|--------------|
| 0 | Hi | greetings |
| 1 | Hi. My name is | greetings |
| 2 | I am facing issues with my credit card | others |
| 3 | please help waive my annual membership | Annual Fee Reversal |
| 4 | please reverse my annual charges | Annual Fee Reversal |

***Figure 5-2.*** *The data*

The Final intent column is the intent of the user. Each intent has a distinct workflow. In this example, you will show the next question following an intent. See Listing 5-2 and Figure 5-3.

***Listing 5-2.***

```
t1["Final intent"].value_counts()
```

```
greetings                  1525
Foreign Travel              700
Card Delivery               600
Report Fraud                600
branch_atm_locator          550
Track application           450
Credit Limit related        400
Pin related                 400
Card Activation             350
Blocked Card                350
Card Cancellation           350
Payment related             300
others                      227
Statement related           150
Annual Fee Reversal         150
Name: Final intent, dtype: int64
```

***Figure 5-3.***

The idea is that once the user's intent is recognized, the bot will ask the next set of guiding questions to resolve the user query. A sample of a workflow for a "reset PIN" is given in Figure 5-4.

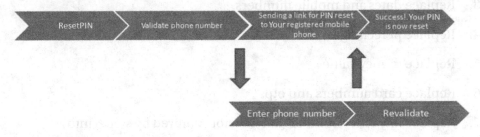

***Figure 5-4.*** *Workflow of classic approach*

For each of the intents, you have the next response to represent the beginning of the workflow to illustrate the concept. See Listing 5-3 and Figure 5-5.

***Listing 5-3.***

```
fp_qns = pd.read_csv('follow_up_qns_v1.csv')
fp_qns.head()
```

| | cat | count | sent |
|---|---|---|---|
| 0 | greetings | 1525 | Hi how can I help you |
| 1 | Foreign Travel | 700 | I understad you want to notify us on your trav... |
| 2 | Card Delivery | 600 | I will quickly check on your card delivery status |
| 3 | Report Fraud | 600 | Please contact XXXX-XXXXX for reporting fraud |
| 4 | branch_atm_locator | 550 | Please provide your address |

***Figure 5-5.***

You will now do some regex preprocessing followed by a `train` and `test` split. You do the following set of operations using regex:

1.  Replacing special characters like \r and \n with a space

2.  Replacing weblinks with a word that indicates that they are URLs. Whenever you replace entities such as these, you add _pp. Hence weblinks get replaced with a token named `url_pp`. This is done in order to indicate it's a replaced word for later analysis. This distinction also helps the model differentiate from a word that is naturally present in the corpus (URL, for instance) to the one replaced using the preprocessing stage (`url_pp`).

3.  Replace dates and mobile numbers.

4.  Replace percentages.

5.  Replace money values.

6.  Replace card numbers and otp.

7.  Replace any other number that was not captured by step 6 into `simp_digit_pp`.

8.  Given that you converted all numbers and special characters so far, you treat any non-digit (other than space or "_") as a space. See Listing 5-4.

***Listing 5-4.***

```python
def preproc(newdf):
    newdf.ix[:,"line1"]=newdf.Line.str.lower()
    newdf.ix[:,"line1"] = newdf.line1.str.replace("inr|rupees","rs")
    newdf.ix[:,"line1"] = newdf.line1.str.replace("\r"," ")
    newdf.ix[:,"line1"] = newdf.line1.str.replace("\n"," ")
    newdf.ix[:,"line1"] = newdf.line1.str.replace("[\s]+"," ")

    newdf.ix[:,"line1"] = newdf.line1.str.replace('http[0-9A-
Za-z:\/\/\.\?\=]*',' url_pp ')
    newdf.ix[:,"line1"] = newdf.line1.str.replace('[0-9]+\/[0-9]+\/
[0-9]+',' date_pp ')
    newdf.ix[:,"line1"] = newdf.line1.str.replace('91[7-9][0-9]{9}',
' mobile_pp ')
    newdf.ix[:,"line1"] = newdf.line1.str.replace('[7-9][0-9]{9}',
' mobile_pp ')

    newdf.ix[:,"line1"] = newdf.line1.str.replace('[0-9]+%',
' digits_percent_pp ')
    newdf.ix[:,"line1"] = newdf.line1.str.replace('[0-9]+percentage',
' digits_percent_pp ')
    newdf.ix[:,"line1"] = newdf.line1.str.replace('[0-9]+th',
' digits_th_pp ')
    newdf.ix[:,"line1"] = newdf.line1.str.replace('rs[., ]*[0-9]+[,.]?
[0-9]+[,.]?[0-9]+[,.]?[0-9]+[,.]?',' money_digits_pp ')
    newdf.ix[:,"line1"] = newdf.line1.str.replace('rs[., ]*[0-9]+',
money_digits_small_pp ')

    newdf.ix[:,"line1"] = newdf.line1.str.replace('[0-9]+[x]+[0-9]*','
cardnum_pp ')
    newdf.ix[:,"line1"] = newdf.line1.str.replace('[x]+[0-9]+',' cardnum_pp ')
    newdf.ix[:,"line1"] = newdf.line1.str.replace('[0-9]{4,7}','
simp_digit_otp ')
    newdf.ix[:,"line1"] = newdf.line1.str.replace('[0-9]+',' simp_digit_pp ')
```

```
newdf.ix[:,"line1"] = newdf.line1.str.replace("a/c"," ac_pp ")
newdf.ix[:,"line1"] = newdf.line1.str.replace('[^a-z _]',' ')

newdf.ix[:,"line1"] = newdf.line1.str.replace('[\s]+,'," ")
newdf.ix[:,"line1"] = newdf.line1.str.replace('[^A-Za-z_]+', ' ')

return newdf
```

```
t2 = preproc(t1)
```

Next, you do a stratified sampling of the train and test datasets. train is used for building the model and test is done for validation. Stratified sampling is used to keep the distribution of the 14 intents similar in train and test. See Listing 5-5.

***Listing 5-5.***

```
tgt = t2["Final intent"]

from sklearn.model_selection import StratifiedShuffleSplit
sss = StratifiedShuffleSplit(test_size=0.1,random_state=42,n_splits=1)

for train_index, test_index in sss.split(t2, tgt):
    x_train, x_test = t2[t2.index.isin(train_index)], t2[t2.index.isin
    (test_index)]
    y_train, y_test = t2.loc[t2.index.isin(train_index),"Final intent"],
    t2.loc[t2.index.isin(test_index),"Final intent"]
```

# Quick Overview of LSTM

Unlike classic data mining models, text features are related to each other in a sequence. For instance, if the first word of a sentence is "I," it will likely be followed by "am." The word "am" will in all probability be followed by a name or a verb. Hence, treating the words as if they are unrelated to each other in a TF-IDF matrix does not do full justice to the problem at hand. Ngrams can solve for some of the sequences but longer ngrams create a sparse matrix. Hence, for this problem you will solve it as a sequence-to-sequence classifier.

Recurrent neural networks help us solve the sequence-to-sequence problem where sentences are treated as sequences. Figure 5-6 shows a simple architecture of the network from https://medium.com/towards-artificial-intelligence/ introduction-to-the-architecture-of-recurrent-neural-networks-rnns-a277007984b7.

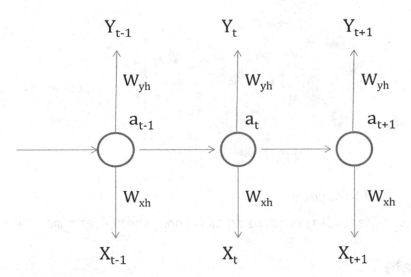

*Figure 5-6.* *RNN architecture*

Here the hidden layer output of xt-1 outputs to the next layer along with the next time input $X_t$. Similarly, the third layer takes hidden layer input of $X_t$ and X. However, this architecture in Figure 5-6 does not know which inputs to "remember" and which inputs to "forget." In a sentence, not every word impacts the probability of the occurrence of other words. For example, in "She visits the mall often, the dependence of "often" is high on the word "visits" as compared to the words "she" or "mall." You also need a mechanism to let the sequential network know which of the sequences is important and which is better forgotten by the network. Also, RNNs are found to suffer from the vanishing gradient problem: as the number of time components extends, the gradient during the backpropagation stage starts to diminish and practically cannot contribute anything to the learning algorithm. You can learn about the vanishing gradient problem in detail at `www.superdatascience.com/blogs/recurrent-neural-networks-rnn-the-vanishing-gradient-problem/`.

In order to solve these problems, LSTM was proposed by Hochreiter and Schmidhuber. LSTM is quite popular nowadays to be used in all kinds of sequence to sequence problems - be it in NLP like text generation or in time sensitive forecasting models. At the basic level, LSTM consists of a cell that outputs two values: a hidden state and a cell state. The cell state is the memory of each unit of LSTM that gets manipulated by the three gates to help the cell understand what to forget and what extra information to add. The hidden state is the modified cell state that is useful as an output from the cell. The states are regulated by three gates: the forget gate, the input gate, and the output gate. See Figure 5-7.

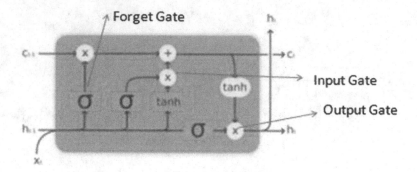

***Figure 5-7.*** *The LSTM gates*

*Source:* `https://en.wikipedia.org/wiki/Long_short-term_memory`

## Forget Gate

The forget gate modifies the cell state value to forget some part of the information. This is done by merging the previous hidden state with the current value and squishing their values using a sigmoid function. This array of 0 to 1 numbers acts as a forget gate when multiplied with the cell states of the last cell. See Equation 5-1 and Figure 5-8.

$$f_t = \sigma_g\left(W_f x_t + U_f h_{t-1} + b_f\right)$$         (Equation 5-1)

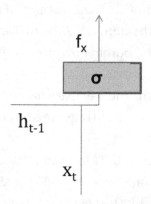

***Figure 5-8.*** *The forget gate*

*Source:* *(Analytics Vidhya)* `www.analyticsvidhya.com/blog/2017/12/ fundamentals-of-deep-learning-introduction-to-lstm/a`

# Input Gate

The input gate finds out what information needs to be added to the cell state. It's a weighted multiplication of the last hidden and the current input with sigmoid and tan activations. First, the hidden state of the last cell gets multiplied with current values and squashed by a sigmoid function. The hidden state of the last cell gets multiplied with current cell values but squashed by a tan function. These two vectors then get multiplied, to which the cell value is added. Unlike the forget gate, which multiplied values ranging from 0 to 1 to the previous cell state, the input gate is additive and adds the output of the set of operations with the cell state. See Equation 5-2.

$$i_t = \sigma_g\left(W_i x_t + U_i h_{t-1} + b_i\right)$$
$$\tilde{c}_t = \sigma_h\left(W_c x_t + U_c h_{t-1} + b_c\right)$$

(Equation 5-2)

The final gate is a product of $c_t$ and $i_t$. This then gets added to the modified cell state. See Equation 5-3 and Figure 5-9.

$$c_t = f_t \circ c_{t-1} + i_t \circ \tilde{c}_t$$

(Equation 5-3)

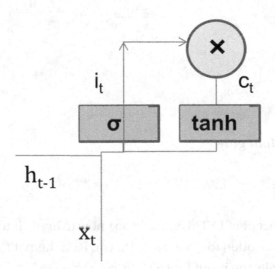

***Figure 5-9.*** *The input gate*

## Output Gate

The output gate determines what can be taken as an output from the cell state. Not all values are relevant in a cell state to be consumed as an output from that cell. The output state is a multiplication of the hidden state of the last cell and the current output. This becomes the filter when squished with a sigmoid function. The cell state you have so far ($C_t$) is squished into a tan function. This cell state is then modified by the filter you have. See Equation 5-4.

$$o_t = \sigma_g \left( W_o x_t + U_o h_{t-1} + b_o \right)$$
$$h_t = o_t \circ \sigma_h \left( c_t \right)$$

(Equation 5-4)

This $h_t$ is the output from the cell. See Figure 5-10.

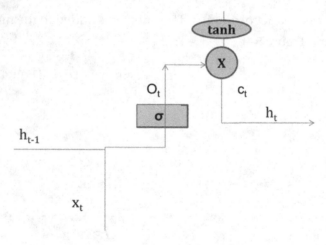

*Figure 5-10.*  *The output gate*

## Applying LSTM

Having learned the concept of LSTM, you will see how to apply it to your problem at hand. With sentences encoded to a fixed length, you feed them it to the classifier to get the final intents. Typically the input for LSTM is converted to N * L * V dimensions.

N refers to the number of rows. L is the maximum length of the sentence. Different sentences are of different lengths and you want all the sentences to be the same length. This is done using padding. You can either cut a sentence to only the first few X words or the last X few words that are important to your problem or you can take the maximum length of a sentence in a given corpus as the L value. So if a sentence is of only 5 words and you determine the L value to be 10, you would pad the sentence with some arbitrary word for the rest of the length. Finally, V is the vocabulary of the corpus and you convert this to one hot encoding.

However, in your case, you are going to feed the network with padded sentences and use an embedding layer to convert the shape LSTM needs. The output of the embedding (without flattening) is fed to the LSTM layers. You are interested in getting the input to the N*L format.

The function in Listing 5-6 finds the maximum length of the sentence (L). This is followed by a function for converting texts to numbers and then padding each of them to the required length.

***Listing 5-6.***

```
def get_max_len(list1):
    len_list = [len(i) for i in list1]
    return max(len_list)
```

You use Keras for the exercise. The installation of Keras and TensorFlow was covered in the last chapter. See Listings 5-7 and 5-8.

***Listing 5-7.***

```
from keras.preprocessing.text import Tokenizer
from keras.preprocessing.sequence import pad_sequences
from keras.utils import to_categorical
tokenizer = Tokenizer()

def conv_str_cols(col_tr,col_te):

    tokenizer = Tokenizer(num_words=1000)
    tokenizer.fit_on_texts(col_tr)

    col_tr1 = tokenizer.texts_to_sequences(col_tr)
    col_te1 = tokenizer.texts_to_sequences(col_te)
    max_len1 = get_max_len(col_tr1)
    col_tr2 = pad_sequences(col_tr1, maxlen=max_len1, dtype='int32',
    padding='post')
```

```
    col_te2 = pad_sequences(col_te1, maxlen=max_len1, dtype='int32',
    padding='post')
    return col_tr2,col_te2,tokenizer,max_len1
```

***Listing 5-8.***

```
tr_padded,te_padded, tokenizer,max_len1 = conv_str_cols(x_train["line1"],
x_test["line1"])
```

```
tr_padded.shape,te_padded.shape
((6391, 273), (711, 273))
```

You have in this array N * L (length with padding). You now convert the dependent variable into the one hot encoded format so that you can apply a softmax classifier at the end. See Listings 5-9 and 5-10.

***Listing 5-9.***

```
From keras.utils import to_categorical
from sklearn.preprocessing import LabelEncoder
le = LabelEncoder()
y_train1 = le.fit_transform(y_train)
y_test1 = le.fit_transform(y_test)

y_train2 = to_categorical(y_train1)
y_test2 = to_categorical(y_test1)
```

***Listing 5-10.***

```
classes_num = len(y_train.value_counts())
```

Your architecture is simple. You embed the input layers and pass them to LSTM. The output of LSTM gets fed to a time-distributed dense layer. This layer gets fed into a fully connected dense layer with classes_num as the number of nodes. One point to note with the number of nodes in LSTM: They are the equivalent of the hidden nodes in a feed-forward neural network. Figure 5-11 quickly explains this and it refers to https://jasdeep06.github.io/posts/Understanding-LSTM-in-Tensorflow-MNIST/.

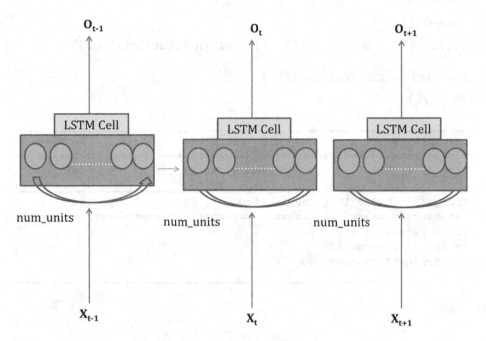

*Figure 5-11.* *Interpreting LSTM and num_units*

# Time-Distributed Layer

The output of the LSTM feeds to a time-distributed layer. What this means is each time the output of the LSTM feeds to a separate dense layer and you extract n time step value for each of the LSTM time inputs. For example, for an LSTM with 5 hidden nodes and 3 time-steps as input, the input to the time distributed layer with node 1 is 3d - layer (Samples * 3 * 5) and the output is 3d-layer (samples * 3* 1). Basically you are extracting one value per time step from the LSTM sequence. This is further fed into dense layers.

For example, if you define a time-distributed layer with 5 nodes, then the output is of shape samples * 3* 5. Let's see a small example to demonstrate this fact with a 5 input time step as input and 6 nodes in a time-distributed layer. The `return_sequences` parameter returns the states at each time step. See Listing 5-11 and Figure 5-12.

*Listing 5-11.*

```
from keras.models import Sequential
from keras.layers import Dense
from keras.layers import Flatten
from keras.layers.embeddings import Embedding
from keras.layers import LSTM,Flatten,TimeDistributed
```

```
model = Sequential()
model.add(LSTM(5, input_shape=(3, 1), return_sequences=True))

model.add(TimeDistributed(Dense(6)))
model.summary()
```

```
Layer (type)                    Output Shape          Param #
=================================================================
lstm_43 (LSTM)                  (None, 3, 5)            140

time_distributed_18 (TimeDis (None, 3, 6)              36
=================================================================
Total params: 176
Trainable params: 176
Non-trainable params: 0
```

***Figure 5-12.***

Let's now get back to the original classifier code. You take an embedded input here and stack the first input (outputs of all the input timesteps) to the second LSTM. The second LSTM with 50 hidden units provides a 3-D input to the time-distributed layer. The time-distributed layer output gets passed to the final dense layer. All the time step nodes (50) share the same weights. A softmax classification is then performed. See Listing 5-12 and Figure 5-13.

***Listing 5-12.***

```
from keras.models import Sequential
from keras.layers import Dense
from keras.layers import Flatten
from keras.layers.embeddings import Embedding
from keras.layers import LSTM,Flatten,TimeDistributed

model = Sequential()
model.add(Embedding(1000, 100, input_length=max_len1))
model.add(LSTM(100,return_sequences=True))
model.add(LSTM(50,return_sequences=True))

model.add(TimeDistributed(Dense(50, activation='relu')))
model.add(Flatten())
```

```
model.add(Dense(classes_num, activation='softmax'))
# compile the model
model.compile(optimizer='adam', loss='categorical_crossentropy',
metrics=['accuracy'])
# summarize the model
print(model.summary())
# fit the model
model.fit(tr_padded, y_train2, epochs=10, verbose=2,batch_size=30)
```

```
Layer (type)                 Output Shape          Param #
=================================================================
embedding_24 (Embedding)     (None, 273, 100)      100000

lstm_37 (LSTM)               (None, 273, 100)      80400

lstm_38 (LSTM)               (None, 273, 50)       30200

time_distributed_13 (TimeDis (None, 273, 50)       2550

flatten_10 (Flatten)         (None, 13650)         0

dense_30 (Dense)             (None, 15)            204765
=================================================================
Total params: 417,915
Trainable params: 417,915
Non-trainable params: 0
```

*Figure 5-13.*

Now you get accuracies with respect to your test dataset. See Listing 5-13.

*Listing 5-13.*

```
pred_mat = model.predict_classes(te_padded)
from sklearn.metrics importaccuracy_score
print (accuracy_score(y_test1, pred_mat))
from sklearn.metrics import f1_score
print (f1_score(y_test1, pred_mat,average='micro'))
print (f1_score(y_test1, pred_mat,average='macro'))
```

```
0.9985935302390999
0.9985935302390999
0.9983013632525033
```

The accuracies are pretty high since it's a demo dataset. Now you proceed to use the intent and get the next response from the bot using the intent.

You use the fp_qns dataset to get the right response for the right intent. First, when a new sentence is posted to the bot, you preprocess and then convert it to a sequence and pad it to the maxlen. You then predict and find out the right response from the fp_qns dataset. For an example, see Listings 5-14 and 5-15 and Figure 5-14.

***Listing 5-14.***

```python
def pred_new_text(txt1):
    print ("customer_text:",txt1)

    newdf = pd.DataFrame([txt1])
    newdf.columns = ["Line"]
    newdf1 = preproc(newdf)

    col_te1 = tokenizer.texts_to_sequences(newdf1["line1"])
    col_te2 = pad_sequences(col_te1, maxlen=max_len1, dtype='int32',
    padding='post')

    class_pred = le.inverse_transform(model.predict_classes(col_te2))[0]

    resp = fp_qns.loc[fp_qns.cat==class_pred,"sent"].values[0]

    print ("Bot Response:",resp,"\n")
    return
```

***Listing 5-15.***

```python
pred_new_text("want to report card lost")
pred_new_text("good morning")
pred_new_text("where is atm")
pred_new_text("cancel my card")
```

```
customer_text: want to report card lost
Bot Response: Please contact XXXX-XXXXX for reporting fraud

customer_text: good morning
Bot Response: Hi how can I help you

customer_text: where is atm
Bot Response: Please provide your address

customer_text: cancel my card
Bot Response: Please contact XXXX-XXXXX for cancellation queries
```

*Figure 5-14.*

# Approach 2 - The Generating Responses Approach

The previous approach is restrictive and custom-made for a business process. You need to be able to classify intents and then match them to the next response. This is very important for the success of the last approach. Let's assume a case in which you are provided with a large corpus of conversations between a customer and a customer service agent. Instead of creating a training sample to map the intent and then to the response, you could directly use the conversation corpus to train for question-answer pairs. A very popular architecture called encoder-decoder is commonly used to train a conversational corpus.

## Encoder-Decoder Architecture

At the heart of the encoder-decoder architecture are LSTMs that take text sequence data as input and provide sequence data as output. Fundamentally you are encoding the sentences of the question using a set of LSTMs and another set of LSTMs to decode the response. The architecture is shown in Figure 5-15.

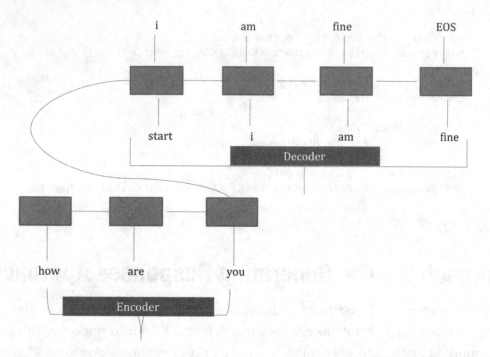

***Figure 5-15.*** *Encoder-decoder architecture*

*Source:  Sequence to Sequence Learning with Neural Networks,IlyaSutskever and Oriol Vinyals and Quoc V. Le, 2014*

Each rectangle box represents a LSTM unit that takes in sequence-based input. As the name implies, the architecture consists of two units: the encoder and decoder.

- **Encoder**: It takes in the input data of questions into the LSTM units and outputs an encoder vector. The encoder vector consists of hidden vectors and outputs cell states of dimensions equal to the hidden dimensions.

- **Decoder**: The decoder gets initialized with the initial encoder states. The decoder input is first appended with Start- and End-of-sentence (EOS) tokens at the beginning and the end, respectively. The architecture uses a technique known as "teacher forcing." During training, the actual values of the decoder are fed into the LSTMs instead of the output from the last step. During inference, the previous timestep output is used to get the output from the decoder. In the architecture in Figure 5-15, the decoder inputs have shifted one timestep and the output of the decoder is the original output. This is to say that the decoder output at time t is the decoder input at time t-1.

# Dataset

For this exercise, you are going to use the dialog dataset `Lets Go` bus dataset. The details of the dataset can be found at `https://dialrc.github.io/LetsGoDataset/`. Let's Go was funded by the National Science Foundation. The Let's Go public spoken dialog system provides bus schedule information to the Pittsburgh population during off-peak times. The data consists of multiple turns of conversation between the user and the voice bot. An example is given in Table 5-2.

*Table 5-2.* *Let's Go Dataset Sample*

| user_id | bot_text | cust_text |
|---|---|---|
| 2061123000 | welcome to the cmu let's go bus information system. to get help at any time, just say help or press zero. what can i do for you? | penn at butler time is it is the centre at highland from runs the 61a eight thirty 61b can't oakland |
| 2061123000 | leaving from oakland . is this correct? | oakland |
| 2061123000 | leaving from oakland . is this correct? | yes |
| 2061123000 | leaving from oakland . is this correct? right. where do you want to go? | squirrel hill of fifth avenue |
| 2061123000 | going to fifth avenue . is this correct? | yes |
| 2061123000 | going to fifth avenue . is this correct? alright when do you wanna travel? | sunday |
| 2061123000 | at what time do you wanna travel? | at five four very about eleven o'clock |
| 2061123000 | leaving at 11 a.m. . is this correct? | yes |

Note that since this is auto-transcribed from voice to text, some of the customer data is incoherent. You will use this conversation corpus and you want to generate the next response given a customer question. You will be using the encoder-decoder architecture for this purpose.

# Implementing the Architecture

As a first step, you preprocess the customer and bot_text. See Listing 5-16.

***Listing 5-16.***

```python
import pandas as pd
t2 = pd.read_csv("letsgo_op.csv")
```

You replace some common sentences in the bot_text. See Listing 5-17.

***Listing 5-17.***

```python
string_repl = "welcome to the cmu let's go bus information system. to get
help at any time, just say help or press zero. what can i do for you? i am
an automated spoken dialogue system that can give you schedule information
for bus routes in pittsburgh's east end. you can ask me about the following
buses: 28x, 54c, 56u, 59u, 61a, 61b, 61c, 61d, 61f, 64a, 69a, and 501. what
bus schedule information are you looking for? for example, you can say,
when is the next 28x from downtown to the airport? ori'd like to go from
mckeesport to homestead tomorrow at 10 a.m."
t2["bot_text"] = t2["bot_text"].str.replace("welcome.*mckeesport to
homestead.*automated spoken dialogue.*","greeting_templ")
t2["trim_bot_txt"] = t2["bot_text"].str.replace('goodbye.*','goodbye')
```

Given that this data is about bus schedules, customers will often use bus stops and place names. I took the bus stop names of Pittsburg and added few more names to the list like downtown, malls, colleges, etc. You can use this ("bus_stops.npz") to replace some of the bus stop names with some common names to help in generalization. See Listing 5-18.

***Listing 5-18.***

```python
import numpy as np
bus_sch = np.load("bus_stops.npz",allow_pickle=True)
bus_sch1 = list(bus_sch['arr_0'])
bus_sch1
['freeporroad',
 'springarden',
```

```
'bellevue',
'shadeland',
'coraopolis',
'fairywood',
'banksville',
'bowehill',
'carrick',
'homesteaparlimited',
'hazelwood',
'nortbraddock',
'lawrenceville-waterfront',
'edgewootowcenter',
'hamilton',
```

Now you replace the place names in cust_text and bot_text using the bus_sch1 list with the function in Listing 5-19, repl_place_bus, using fuzz.ratio between words in the sentences and bus_sch1. The fuzz.ratio calculates the similarity between strings using Levenshtein Distance (minimum number of single character edits required to change one word to another). fuzzywuzzy is the package used to calculate fuzz.ratio. You can pip install fuzzywuzzy. Also see Listing 5-20.

### Listing 5-19.

```
!pip install fuzzywuzzy
```

### Listing 5-20.

```
from fuzzywuzzy import fuzz
str_list = []
def repl_place_bus(row,req_col):
    ctext = row[req_col]
    fg=0
    str1=ctext

    try:
        for j in ctext.split():
```

```
                    for k in bus_sch1:
                        score = fuzz.ratio(j,k )
                        if(score>=70):
                            #print (i,j,k)
                            fg=1
                            break;
                    #if(fg==1):
                        #break;
                    if(fg==1):
                        fg=0
                        str1 = str1.replace(j," place_name ")

        except:
            print (j,i)
    #str_list.append(str1)

    return str1
```

You have 7000+ rows. In each row, you are iterating through the words in the sentences. This will increase the runtime -quite a bit. In order for it to run faster, you apply the functions on parallelization in Python. You run the function simultaneously across multiple processes using a library called joblib. The n_jobs parameter provides the number of processes that are going to be run in parallel.

You install the joblib library using pip install. See Listings 5-21 and 5-22.

***Listing 5-21.***

```
!pip install joblib
```

***Listing 5-22.***

```
from joblib import Parallel, delayed
com0 = Parallel(n_jobs=12, backend="threading",verbose=1)(delayed
(repl_place_bus)(row,"cust_text") for i,row in t2.iterrows())
bot_list = Parallel(n_jobs=12, backend="threading",verbose=1)(delayed
(repl_place_bus)(row,"trim_bot_txt") for i,row in t2.iterrows())
t2["corrected_cust"] = com0
t2["corrected_bot"] = bot_list
```

The columns `corrected_cust` and `corrected_bot` contain the replaced sentences. So now you do the following preprocesses: normalizing directions (since it's a bus booking), normalizing A.M./P.M., replacing number words with numbers, normalizing bus names (like 8a,11c etc), time, replacing numbers. See Listings 5-23 and 5-24.

***Listing 5-23.***

```
direct_list = ['east','west','north','south']
def am_pm_direct_repl(t2,col):
    t2[col] = t2[col].str.replace('p.m','pm')
    t2[col] = t2[col].str.replace('p m','pm')
    t2[col] = t2[col].str.replace('a.m','am')
    t2[col] = t2[col].str.replace('a m','am')

for i in direct_list:
        t2[col] = t2[col].str.replace(i,' direction ')

return t2
```

***Listing 5-24.***

```
t3 = am_pm_direct_repl(t2,"corrected_cust")
t3 = am_pm_direct_repl(t2,"corrected_bot")
```

The following function using the num2words package gets numeric numbers for number-words like *one, two, three,* etc. You use this to replace the sentences with the right numbers. You will be replacing "th" words, so for example, Fifth Avenue gets replaced with "num_th." You install the num2words package in Listing 5-25 and use it in Listings 5-26 and 5-27.

***Listing 5-25.*** Installing num2words

```
!pip install num2words
```

***Listing 5-26.*** Using num2words

```
from num2words import num2words
word_num_list = []
word_num_th_list = []
```

209

```
for i in range(0,100):
    word_num_list.append(num2words(i))
    word_num_th_list.append(num2words(i,to='ordinal'))

print (word_num_list[0:5])
print (word_num_th_list[0:5])
['zero', 'one', 'two', 'three', 'four']
['zeroth', 'first', 'second', 'third', 'fourth']
```

***Listing 5-27.***

```
def repl_num_words(t2,col):
    for num,i in enumerate(word_num_th_list):
        t2[col] = t2[col].str.replace(i,"num_th")

        t2[col] = t2[col].str.replace(word_num_list[num],str(num))
    return t2

t4 =  repl_num_words(t3,"corrected_cust")
t4 =  repl_num_words(t3,"corrected_bot")
```

The code in Listing 5-28 replaces bus and time names herewith.

***Listing 5-28.***

```
t4["corrected_cust"] = t4["corrected_cust"].str.replace('[0-9]{1,2}[a-z]
{1,2}','bus_name')
t4["corrected_cust"] = t4["corrected_cust"].str.replace('[0-9]{1,2}
[\.\s]*[0-9]{1,2}[\s]*pm','time_name')
t4["corrected_cust"] = t4["corrected_cust"].str.replace('[0-9]{1,2}
[\.\s]*[0-9]{1,2}[\s]*am','time_name')

t4["corrected_bot"] = t4["corrected_bot"].str.replace('[0-9]{1,2}[a-z]
{1,2}','bus_name')
t4["corrected_bot"] = t4["corrected_bot"].str.replace('[0-9]{1,2}
[\.\s]*[0-9]{0,2}[\s]*pm','time_name')
t4["corrected_bot"] = t4["corrected_bot"].str.replace('[0-9]{1,2}
[\.\s]*[0-9]{0,2}[\s]*am','time_name')
```

Next, you replace special characters and split the bot sentences with a sentence separator, sep_sent. This will help you to templatize each of the bot sentences and generate these templates from the decoder later. See Listing 5-29.

***Listing 5-29.***

```
t4["corrected_bot1"] = t4["corrected_bot"].str.replace('\xa0','')
t4["corrected_bot1"] = t4["corrected_bot1"].str.replace('[^a-z0--
9\s\.\?]+','')
t4["corrected_bot1"] = t4["corrected_bot1"].str.replace('[0-9]{1,3}','
short_num ')
t4["corrected_bot1"] = t4["corrected_bot1"].str.replace('[0-9]{4,}','
long_num ')

t4["corrected_bot1"] = t4["corrected_bot1"].str.replace('(\.\s){1,3}','
sep_sent ')
t4["corrected_bot1"] = t4["corrected_bot1"].str.replace('(\.){1,3}','
sep_sent ')
t4["corrected_bot1"] = t4["corrected_bot1"].str.replace('\?',' sep_sent ')
t4["corrected_bot1"] = t4["corrected_bot1"].str.replace('(sep_sent )+','
sep_sent ')
```

Replacing special characters in customer sentences is done in order to decrease the sparsity of the data. For example, consider two sentences: "Is this correct…" and "Is this correct". The word "correct" is same except in both cases, except one has three dots and the other has none. When you tokenize, you want to make sure that these two words are treated the same. See Listing 5-30.

***Listing 5-30.***

```
t4["corrected_cust"] = t4["corrected_cust"].str.replace('(\.\s){1,3}',' ')
t4["corrected_cust"] = t4["corrected_cust"].str.replace('(\.){1,3}',' ')
```

Since you have a corpus that is an interaction between a bot and customer, some of the sentences are repetitive across different conversations. Given the nature of customer service data, even if it was a human-to-human conversation, there would be a good amount of repetitive sentences used across different dialogues. These repetitive sentences can be converted to templates. For example, when asking about a bus arrival time, there are only a certain number of ways to state the arrival time of a bus. This can greatly help you to normalize the data and have meaningful outputs. In Listing 5-31, you convert the bot sentences to meaningful template sentences. These templates will be mapped to template numbers. See Figure 5-16.

***Listing 5-31.***

```
templ = pd.DataFrame(t4["corrected_bot1"].value_counts()).reset_index()
templ.columns = ["sents","count"]
templ.head()
```

| | sents | count |
|---|---|---|
| 0 | welcome to the cmu lets go bus information system sep_sent to get help at any time just say help or press short_num sep_sent what can i do for you sep_sent | 315 |
| 1 | you can say when is the next bus when is the previous bus placename a new query or goodbye | 244 |
| 2 | for example you can say when is the next busname from placename to the placename or id like to go from placename to placename tomorrow at timename sep_sent | 176 |
| 3 | welcome to the cmu lets go bus information system sep_sent to get help at any time just say help or press short_num sep_sent what can i do for you sep_sent | 175 |
| 4 | leaving at timename sep_sent did i get that placename | 162 |

***Figure 5-16***

You get all the unique sentences into a list. Next, you remake the list into a dataframe to assign each sentence to a template id. See Listing 5-32 and Figure 5-17.

***Listing 5-32.***

```
df = templ.sents.str.split("sep_sent",expand=True)
sents_all = []
for i in df.columns:
    l1 = list(df[i].unique())
    sents_all = sents_all + l1
sents_all1 = set(sents_all)

sents_all_df = pd.DataFrame(sents_all1)
sents_all_df.columns = ["sent"]
sents_all_df["ind"] = sents_all_df.index
sents_all_df["ind"] = "templ_" + sents_all_df["ind"].astype('str')
sents_all_df.head()
```

| | sent | ind |
|---|---|---|
| 0 | did i get that placename alrightwhere are yo... | templ_0 |
| 1 | did i get that placename alrightwhat is your... | templ_1 |
| 2 | going to pittsburgh placename | templ_2 |
| 3 | it will arrive at talbot placename at numth... | templ_3 |
| 4 | the next busname leaves hawkins placename a... | templ_4 |

*Figure 5-17.*

Now you take the original dataset of t4 and assign different sentences to the template ids. First, you create a dictionary with sentences as the keys and template ids. See Listings 5-33 and 5-34.

*Listing 5-33.*

```
sent_all_df1 = sents_all_df.loc[:,["sent","ind"]]
df_sents = sent_all_df1.set_index('sent').T.to_dict('list')
```

*Listing 5-34.*

```
t4["corrected_bots_sents"] = t4["corrected_bot1"].str.split("sep_sent")
index_list=[]
for i, row in t4.iterrows():
    sent_list = row["corrected_bots_sents"]
    str_index = ""
    for j in sent_list:
        if(len(j)>=3):
            str_index = str_index + " " + str(df_sents[j][0])
    index_list.append(str_index)
t4["bots_templ_list"] = index_list
```

Your training set is of a bot asking a question and a customer saying the answer. But for training, your input will be customer sentence and the response from the bot as the output. So you shift corrected_bot1 to one sentence after. This way the corpus becomes the incoming customer text and the right output from the bot. You shift the user_id to mark the end of the chat. See Listing 5-35 and Figure 5-18.

213

***Listing 5-35.***

```
t4["u_id_shift"] = t4["user_id"].shift(-1)
t4["corrected_bot1_shift"] = t4["corrected_bot1"].shift(-1)
t4["bots_templ_list_shift"] = t4["bots_templ_list"].shift(-1)

t4.loc[t4.u_id_shift!=t3.user_id,"corrected_bot1_shift"] = "end_of_chat"
t4.loc[t4.u_id_shift!=t3.user_id,"bots_templ_list_shift"] = "end_of_chat"

req_cols = ["user_id","corrected_cust","corrected_bot1_shift",
"bots_templ_list_shift"]
t5 = t4[req_cols]

t5.tail()
```

| | user_id | corrected_cust | corrected_bot1_shift | bots_templ_list_shift |
|---|---|---|---|---|
| 7755 | 2070619029 | far | for example you can say placename placename... | templ_104 |
| 7756 | 2070619029 | place_name | going to placename sep_sent did i get tha... | templ_1115 templ_270 |
| 7757 | 2070619029 | yes | going to placename sep_sent did i get tha... | templ_1115 templ_910 |
| 7758 | 2070619029 | next bus | i think you placename the next bus sep_se... | templ_774 templ_934 |
| 7759 | 2070619029 | yes | end_of_chat | end_of_chat |

***Figure 5-18.***

You now pick the output of dictionary df_sents for mapping the templates to the original sentences. In the encoder, you use `template_ids` for the output. Then you remap using the dictionary to print the actual sentences. See Listing 5-36.

***Listing 5-36.***

```
import pickle
output = open('dict_templ.pkl', 'wb')

# Pickle dictionary using protocol 0.
pickle.dump(df_sents, output)
```

Finally, you output `t5` to be taken to the encoder-decoder training process. See Listing 5-37.

***Listing 5-37.***

```
t5.to_csv("lets_go_model_set.csv")
```

# Encoder-Decoder Training

Using the columns `corrected_cust` and `bots_templ_list_shift` as input and output, you can train the encoder-decoder model. Given that you are using a large architecture, you will use Google Colab and GPUs for this exercise. Colaboratory (`https://colab.research.google.com/notebooks/`) is a Google research project created to help disseminate machine learning education and research. It's a Jupyter notebook environment that requires no setup to use and runs entirely in the cloud. You can get started and set up a Colab Jupyter notebook.

You first upload the data in Google Drive. You can then mount the data in Google Colab using the code in Listing 5-38. This loads all the files into the drive. When running the code, you need to enter an authorization code by following a link. See Listing 5-38.

***Listing 5-38.***

```
from google.colab import drive
drive.mount('/content/drive')
```

A screen like the one in Figure 5-19 will appear and you need to enter the authorization code shown in the link. Then see Listing 5-39.

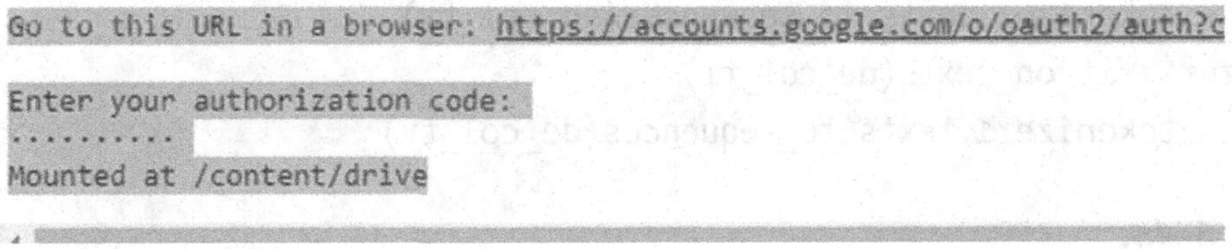

```
Go to this URL in a browser: https://accounts.google.com/o/oauth2/auth?c

Enter your authorization code:
..........
Mounted at /content/drive
```

***Figure 5-19.***  *Authorization for mounting files*

***Listing 5-39.***

```
import pandas as pd
t1 = pd.read_csv(base_fl_csv)
```

You add `start` and `end` tags to the sentences in Listing 5-40.

***Listing 5-40.*** Start and End Tags

```
t1["bots_templ_list_shift"]  = 'start ' + t1["bots_templ_list_shift"] + ' end'
```

You create tokenizers, one for the encoder and one for the decoder, in Listing 5-41.

***Listing 5-41.*** Creating Tokenizers

```
from keras.preprocessing.text import Tokenizer
from keras.preprocessing.sequence import pad_sequences
tokenizer = Tokenizer()
tokenizer1 = Tokenizer()
```

You now tokenize the sentences in the customer and bot text (encoder input and decoder input, respectively) and pad them based on the max length of sentences in the corpus. See Listings 5-42 and 5-43.

***Listing 5-42.***

```
en_col_tr = list(t1["corrected_cust"].str.split())
de_col_tr = list(t1["bots_templ_list_shift"].str.split())

tokenizer.fit_on_texts(en_col_tr)
en_tr1 = tokenizer.texts_to_sequences(en_col_tr)
tokenizer1.fit_on_texts(de_col_tr)
de_tr1 = tokenizer1.texts_to_sequences(de_col_tr)
```

***Listing 5-43.***

```
def get_max_len(list1):
    len_list = [len(i) foriin list1]
return max(len_list)

max_len1 = get_max_len(en_tr1)
max_len2 = get_max_len(de_tr1)

en_tr2 = pad_sequences(en_tr1, maxlen=max_len1, dtype='int32', padding='post')
de_tr2 = pad_sequences(de_tr1, maxlen=max_len2, dtype='int32', padding='post')
de_tr2.shape,en_tr2.shape,max_len1,max_len2
((7760, 27), (7760, 28), 28, 27)
```

You can see that the encoder and decoder input have two-dimensional shapes. Since you are directly feeding them to LSTM (without an embedding layer), you need to convert them into three-dimensional arrays. This is done by converting the sequence of words to one hot encoded forms. See Listings 5-44 and 5-45.

***Listing 5-44.***

```
from keras.utils import to_categorical
en_tr3 = to_categorical(en_tr2)
de_tr3 = to_categorical(de_tr2)
en_tr3.shape, de_tr3.shape
((7760, 28, 733), (7760, 27, 1506))
```

***Listing 5-45.***

```
from keras.utils import to_categorical
en_tr3 = to_categorical(en_tr2)
de_tr3 = to_categorical(de_tr2)
en_tr3.shape, de_tr3.shape
((7760, 28, 733), (7760, 27, 1506))
```

The arrays are now three-dimensional. Please note that so far you have only defined inputs. You must now define the outputs. The output of the model is the decoder with a t +1 timstep. You want to predict the next word, given the last word. See Listing 5-46.

***Listing 5-46.***

```
import numpy as np
from scipy.ndimage.interpolation import shift
de_target3 = np.roll(de_tr3, -1,axis=1)
de_target3[:,-1,:]=0
```

Saving the number of encoder and decoder tokens to define model inputs happens in Listing 5-47.

***Listing 5-47.***

```
num_encoder_tokens = en_tr3.shape[2]
num_decoder_tokens = de_tr3.shape[2]
```

In the encoder-decoder architecture with teacher forcing, there is a difference between the training and inferencing steps. The code is derived from this article at `https://blog.keras.io/a-ten-minute-introduction-to-sequence-to-sequence-learning-in-keras.html`. First, you see the steps for training. Here you define LSTMs with 100 hidden nodes for each step of the encoder-input (customer text). You keep the final cell state and the hidden state of the encoder and discard the outputs from each of the LSTM cells. The decoder takes input from the bot text and initializes the initial state from the encoder output. Since the encoder has 100 hidden nodes, the decoder also has 100 hidden nodes. The output of the decoder at each cell is passed to a dense layer. The network is trained with the time adjusted lag of the decoder input. Basically, the encoder values and decoder inputs at t-1 predict the bot text at t. See Listing 5-48.

***Listing 5-48.***

```python
from keras.models import Model
from keras.layers import Input, LSTM, Dense

encoder_inputs = Input(shape=(None, num_encoder_tokens))
encoder = LSTM(100, dropout=.2,return_state=True)
encoder_outputs, state_h, state_c = encoder(encoder_inputs)
# We discard `encoder_outputs` and only keep the states.
encoder_states = [state_h, state_c]

decoder_inputs = Input(shape=(None, num_decoder_tokens))

decoder_lstm = LSTM(100, return_sequences=True, return_
state=True,dropout=0.2)
decoder_outputs, _, _ = decoder_lstm(decoder_inputs,
                                    initial_state=encoder_states)
decoder_dense = Dense(num_decoder_tokens, activation='softmax')
decoder_outputs = decoder_dense(decoder_outputs)

# Define the model that will turn
# `encoder_input_data` & `decoder_input_data` into `decoder_target_data`
model = Model([encoder_inputs, decoder_inputs], decoder_outputs)
```

You will now see the setup for model inference.

# Encoder Output

First, you prepare a layer to get the encoder states. These are the initial states for the decoder.

# Decoder Input

The decoder input is not known when the model runs. Hence you must use the architecture to predict one word at a time and use that word to predict the next word. The decoder part of the model takes the decoder (time delayed input). For the first word, the decoder model is initialized with encoder states. The decoder_lstm layer is called using decoder inputs and the initialized encoder states. This layer provides two sets of outputs: the output set of the cells (decoder_ouput) and the final output of cell_state and hidden_state. The decoder_ouput is passed to the dense layer to get the final prediction of the bot text. The decoder state's output is used to update the states for the next run (next word prediction). The bot_text is appended to the decoder input and the same process is repeated. See Listing 5-49.

***Listing 5-49.***

```
encoder_model = Model(encoder_inputs, encoder_states)

decoder_state_input_h = Input(shape=(100,))
decoder_state_input_c = Input(shape=(100,))
decoder_states_inputs = [decoder_state_input_h, decoder_state_input_c]
decoder_outputs, state_h, state_c = decoder_lstm(
decoder_inputs, initial_state=decoder_states_inputs)
decoder_states = [state_h, state_c]
decoder_outputs = decoder_dense(decoder_outputs)
decoder_model = Model(
    [decoder_inputs] + decoder_states_inputs,
    [decoder_outputs] + decoder_states)
```

Now you must set up the layers for the model, for training and inference. In Listing 5-50, you train the model.

***Listing 5-50.***

```
from keras.callbacks import EarlyStopping
from keras.callbacks import ModelCheckpoint
from keras.optimizers import Adam

model.compile(optimizer='rmsprop', loss='categorical_crossentropy')
model.fit([en_tr3, de_tr3], de_target3,
          batch_size=30,
          epochs=30,
          validation_split=0.2)
```

You now want to test if the model works, so you save the model objects. Use the relevant root_dir folder here. See Listing 5-51.

***Listing 5-51.***

```
from tensorflow.keras.models import save_model
dest_folder = root_dir + '/collab_models/'
encoder_model.save( dest_folder + 'enc_model_collab_211_redo1')
decoder_model.save( dest_folder + 'dec_model_collab_211_redo1')
import pickle
dest_folder = root_dir + '/collab_models/'
output = open(dest_folder + 'tokenizer_en_redo.pkl', 'wb')

# Pickle dictionary using protocol 0.
pickle.dump(tokenizer, output)

dest_folder = root_dir + '/collab_models/'
output1 = open(dest_folder + 'tokenizer_de_redo.pkl', 'wb')
pickle.dump(tokenizer1, output1)
```

Open a new notebook and write the execution of the inference part. This code is written locally and you get the files from Google Colab. The code has three parts: the text given the preprocessing (replacing appropriate words and regular expressions), the conversion to text to the required array format, and the model inference.

Once you fit the model, you run the inference code one word at a time. The starting word is initialized to the start token of the word. Please note the decoder states reinitialize for the next word at the end of the code block.

First, you import all the saved models into the new session. See Listing 5-52.

**Listing 5-52.**

```
import pandas as pd
from keras.models import Model
from tensorflow.keras.models import load_model
```

Then you load the encoder and decoder models. See Listing 5-53.

**Listing 5-53.**

```
encoder_model = load_model(dest_folder + 'enc_model_collab_211_redo1')
decoder_model = load_model(dest_folder +'dec_model_collab_211_redo1')
```

You load the encoder and decoder tokenizers and the dictionary files for converting templates to sentences. Use the relevant dest_folder. This is where you downloaded the model files from Google Colab. See Listing 5-54.

**Listing 5-54.**

```
import pickle
import numpy as np
pkl_file = open(dest_folder + 'tokenizer_en_redo_1.pkl', 'rb')

tokenizer = pickle.load(pkl_file)

pkl_file = open(dest_folder + 'tokenizer_de_redo.pkl', 'rb')

tokenizer1 = pickle.load(pkl_file)

pkl_file = open( 'dict_templ.pkl', 'rb')

df_sents = pickle.load(pkl_file)

bus_sch = np.load("bus_stops.npz",allow_pickle=True) #make sure you use
the .npz!
bus_sch1 = list(bus_sch['arr_0'])
```

221

# Preprocessing

The preprocessing steps are highlighted in Listing 5-55. (You convert the pandas `replace` to regex `replace`). This is a repeat of the process you followed when training: replacing place names, directions, number names, bus names, times, and special characters. This gets applied to the customer sentences and then fed into the model. Also see Listing 5-56.

***Listing 5-55.***

```
from fuzzywuzzy import fuzz
def repl_place_bus(ctext):
    #for i,row in t2.iterrows():
    #ctext = row[req_col]
    fg=0
    str1=ctext

    for j in ctext.split():
        #print (j)

        for k in bus_sch1:
            score = fuzz.ratio(j,k )
            if(score>=70):
                #print (i,j,k)
                fg=1
                break;
        #if(fg==1):
            #break;
        if(fg==1):
            fg=0
            str1 = str1.replace(j," place_name ")

    return str1
```

***Listing 5-56.***

```
direct_list = ['east','west','north','south']
def am_pm_direct_repl(ctext):
    ctext = re.sub('p.m','pm',ctext)
```

```
    ctext = re.sub('p m','pm',ctext)
    ctext = re.sub('a.m','am',ctext)
    ctext = re.sub('a m','am',ctext)

    for i in direct_list:
        ctext = re.sub(i,' direction ',ctext)

    return ctext

from num2words import num2words
word_num_list = []
word_num_th_list = []
for i in range(0,100):
    word_num_list.append(num2words(i))
    word_num_th_list.append(num2words(i,to='ordinal'))

def repl_num_words(ctext):
    for num,i in enumerate(word_num_th_list):
        ctext = re.sub(i,"num_th",ctext)
        ctext = re.sub(word_num_list[num],str(num),ctext)
    return ctext

def preprocces(ctext):
    ctext = ctext.lower()
    ctext1 = repl_place_bus(ctext)
    ctext1 = am_pm_direct_repl(ctext1)
    ctext1 = repl_num_words(ctext1)

    ctext1 = re.sub('[0-9]{1,2}[a-z]{1,2}','bus_name',ctext1)
    ctext1 = re.sub('[0-9]{1,2}[\.\s]*[0-9]{1,2}[\s]*pm','time_name',
    ctext1)
    ctext1 = re.sub('[0-9]{1,2}[\.\s]*[0-9]{1,2}[\s]*am','time_name',
    ctext1)

    ctext1 =re.sub('(\.\s){1,3}',' ',ctext1)
    ctext1 =re.sub('(\.){1,3}',' ',ctext1)

    return ctext1
```

You now test the preprocess using the code in Listing 5-57.

*Listing 5-57.*

```
sent = "when is the next bus to crafton boulevard 9 00 P.M east. . 85c"
preprocces(sent)
'when is the next bus to craftonplace_nametime_name direction bus_name'
```

The sentence seems to have been preprocessed as intended. You are now set to convert the preprocessed text into the right array of numbers as defined during training using the tokenizer objects. You set the padding length and number of classes for one hot from the values in the training process. See Listings 5-58 and 5-59.

*Listing 5-58.*

```
max_len1 = 28
num_encoder_tokens =733
num_decoder_tokens = 1506
```

*Listing 5-59.*

```
def conv_sent(sent,shp):
    sent = preprocces(sent)
    #print (sent)
    en_col_tr = [sent.split()]

    en_tr1= tokenizer.texts_to_sequences(en_col_tr )

    en_tr2 = pad_sequences(en_tr1, maxlen=max_len1, dtype='int32',
    padding='post')
    en_tr3 = to_categorical(en_tr2,num_classes =shp)

    return en_tr3
```

You check this using your test preprocessed sentence and see if the shape is as expected (rows* length of sequence * one-hot). See Listing 5-60.

*Listing 5-60.*

```
sent1 = 'when is the next bus to craftonplace_nametime_
name  direction  bus_name'
sent1_conv = conv_sent(sent1,num_encoder_tokens)
sent1_conv.shape
(1, 28, 733)
```

Now you apply the model to the converted sentences. As discussed, the first word will be initialized to the start token and the states of the decoder will be initialized to encoder states. With that, you get the next word. Now, with the next word and the states of the word as the set of inputs, you get the word at t+1. You repeat the process until you reach the end token. Remember you trained on bot sentence templates and not on the actual sentences. Hence you get the template id as the output. You then convert the template id to the required word using Listing 5-61.

*Listing 5-61.*

```
def rev_dict(req_word):
    str1=""
    for k,v in df_sents.items():
        if(df_sents[k]==[req_word]):
            str1 = k
    return str1
from keras.preprocessing.sequence import pad_sequences
from keras.utils import to_categorical
import numpy as np

def get_op(text1,shp):

    st_pos = 0
    input_seq = conv_sent(text1, shp)

    states_value = encoder_model.predict(input_seq)
    target_seq = np.zeros((1, 1, num_decoder_tokens))
    str_all = ""
    # Populate the first word of target sequence with the start word.
    target_seq[0, 0, st_pos] = 1
    for i in range(0,10):
```

225

```
        decoded_sentence = ''
        output_tokens, h, c = decoder_model.predict(
                [target_seq] + states_value)
        sampled_token_index = np.argmax(output_tokens[0, -1, :])

        sampled_char = tokenizer1.sequences_to_texts([[sampled_token_index]])
        str1 = rev_dict(sampled_char[0])
        str_all = str_all + "\n" + str1

        if(sampled_char[0]=='end'):
            break;
        else:
            decoded_sentence += sampled_char[0]
            target_seq = np.zeros((1, 1, num_decoder_tokens))
            target_seq[0, 0, sampled_token_index] = 1.
            ####initializing state values
            states_value = [h, c]

    return str_all
```

You can check the output herewith. See Listings 5-62, 5-63, and 5-64.

## Listing 5-62.

```
text1 = "54c"
get_op(text1,num_encoder_tokens)
```

'\n placename going to placename transportation center \n short_numshort_num n short_num where are you leaving from \n'

## Listing 5-63.

```
text1 = "next bus to boulevard"
get_op(text1,shp)
```

'\n did i get that placename okay where would you like to leave from \n if you placename to leave from placename and bausman say yes or press short_num otherwise say no or press short_num \n'

*Listing 5-64.*

```
text1 = "Do you have bus to downtown in the morning"
get_op(text1,shp)

'\n when would you like to travel  \n'
```

You seem to have a start here. The following can be done to improve the responses:

1. You need to replace the answers with the appropriate place names and present a more meaningful answer.

2. You have trained on a single input. But it's a conversation so you need to train it on the n-1 response and the question to get the next response.

3. Some of the transcriptions were not accurate. If they can be weeded out and the model trained with a cleaner corpus, it will be better.

4. You can use embeddings (pretrained or trained during the model) as inputs to LSTMs.

5. You can use tuning hyperparameters like learning rates or learning algorithms.

6. You can get stacked LSTM layers (the first set of LSTMs provides input to the second set of LSTMs).

7. You can use bidirectional LSTMs. Later you'll see an example of a bidirectional LSTM.

8. You can use advanced architectures like BERT.

# Bidirectional LSTM

Take this sentence: "She was playing tennis." The context for the word "playing" is more obvious from "tennis." In order to find the word given a context, not only the words before (past words) but the words ahead of the word of interest (future words) are also important. So you train the model with one forward sequence and one backward sequence. This is the architecture from http://colah.github.io/posts/2015-09-NN-Types-FP/. The output from both these LSTMs can be concatenated or summed. See Figure 5-20.

227

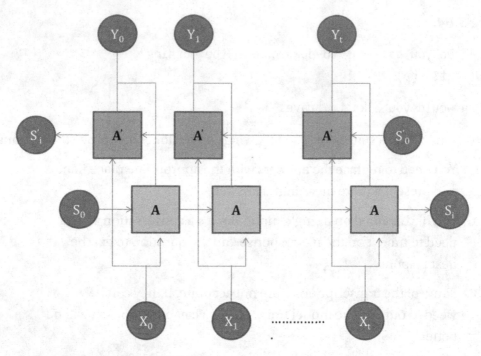

**Figure 5-20.**  *Bidirectional LSTM*

Let's look at a small implementation of the problem at hand using a bidirectional LSTM. Here the outputs are concatenated and passed to the decoder.

## Encoder

Bidirectional LSTMs have additional outputs of forward and backward hidden states (forward_h, forward_c, backward_h, and backward_c). They are then concatenated (hidden states together and cell states together) to get the final encoder states.

## Decoder

The decoder has a unidirectional LSTM, as with the earlier case. However, the number of hidden units is equal to the concatenated hidden units of the bidirectional LSTM of the encoder. In your case (Listing 5-65), the encoder LSTMs have 100 units each and the decoder has 200 units.

*Listing 5-65.*

```
from keras.layers import LSTM,Bidirectional,Input,Concatenate

from keras.models import Model
from keras.models import Model
from keras.layers import Input, LSTM, Dense

n_units = 100
n_input = num_encoder_tokens
n_output = num_decoder_tokens

# encoder
encoder_inputs = Input(shape=(None, n_input))
encoder = Bidirectional(LSTM(n_units, return_state=True))
encoder_outputs, forward_h, forward_c, backward_h, backward_c =
encoder(encoder_inputs)
state_h = Concatenate()([forward_h, backward_h])
state_c = Concatenate()([forward_c, backward_c])
encoder_states = [state_h, state_c]

# decoder
decoder_inputs = Input(shape=(None, n_output))
decoder_lstm = LSTM(n_units*2, return_sequences=True, return_state=True)
decoder_outputs, _, _ = decoder_lstm(decoder_inputs, initial_state=encoder_
states)
decoder_dense = Dense(n_output, activation='softmax')
decoder_outputs = decoder_dense(decoder_outputs)
model = Model([encoder_inputs, decoder_inputs], decoder_outputs)

# define inference encoder
encoder_model = Model(encoder_inputs, encoder_states)
# define inference decoder
decoder_state_input_h = Input(shape=(n_units*2,))
decoder_state_input_c = Input(shape=(n_units*2,))
decoder_states_inputs = [decoder_state_input_h, decoder_state_input_c]
decoder_outputs, state_h, state_c = decoder_lstm(decoder_inputs, initial_
state=decoder_states_inputs)
```

```
decoder_states = [state_h, state_c]
decoder_outputs = decoder_dense(decoder_outputs)
decoder_model = Model([decoder_inputs] + decoder_states_inputs, [decoder_
outputs] + decoder_states)
```

The encoder-decoder architecture is a common way of handling sequence-to-sequence problems.

# BERT

BERT (Bidirectional Encoder Representations from Transformers) is the new paradigm in NLP. It is getting widely adopted in sequence-to-sequence modelling tasks as well as in classifiers. BERT is a pretrained model. It was released by Google in 2018 and has been found to outperform in various NLP challenges since then. The commercial use of BERT is also catching up. Many organizations are fine-tuning their context-specific use cases in NLP using the base BERT model.

## Language Models and Fine-Tuning

Like the imagenet moment of image processing, BERT is the turning point in NLP. In text mining problems we are always limited by the training dataset. Add to it the complexity of language, where words have similar meanings, the usage of words in various contexts, mixing words from different languages in the same sentences, etc. To address these problems (sparsity of data and complexity of language), two concepts have evolved over the last few years: language models and fine tuning. Language models are very useful. They capture the semantic meanings of a language. You learned about word2vec in the last chapter, where the meanings of the words are captured using pretrained models. BERT is also a powerful language model where the meaning of a word is captured along with its context. To quote a familiar example, BERT can differentiate the word "bank" between the two sentences "She went to the bank to deposit money." and "Coconuts grow well in the river bank where the soil is rich." BERT helps us to capture the disambiguation of the same word, "bank," based on the words preceding and succeeding it. Please note that in word2vec models, they are not distinguishable.

Image classifications have been primarily using fine-tuning methods for some time now. NLP has started adopting fine tuning of late. Basically a large corpus of text data is pretrained using neural networks on a large sample with millions of training datasets. This trained neural network, along with its weights, are open sourced. A NLP scientist can now use this pretrained model for their downstream tasks like classification, summarization, etc. for a specific context. They can retrain the context-specific models with new data but initialize with the weights relevant to the pretrained models. BERT is a language model pretrained and available for fine tuning for downstream NLP tasks. Let's see an example using BERT for the classification of a credit card company's dataset into multiple intents.

## Overview of BERT

BERT architecture on the outset is built on transformers. BERT is also bidirectional. The words take contexts from words preceding and succeeding in the sentences. Before you go further, you need to understand transformers. In the sentence "The animal didn't cross the road because it was afraid," the word "it" refers to the animal. In the sentence "The animal didn't cross the road because it was narrow," the word "it" refers to the road. (Example from `https://ai.googleblog.com/2017/08/transformer-novel-neural-network.html`). As you can see, the surrounding words determine the meaning of the word. This brings up the concept of attention, which is at the core of the transformers. Let's see the architecture of transformers. This is from the paper "Attention is all you need" at `https://arxiv.org/abs/1706.03762`. Figure 5-21 shows a single set of encoder-decoder. In the paper, the author shows six sets of encoder-decoder.

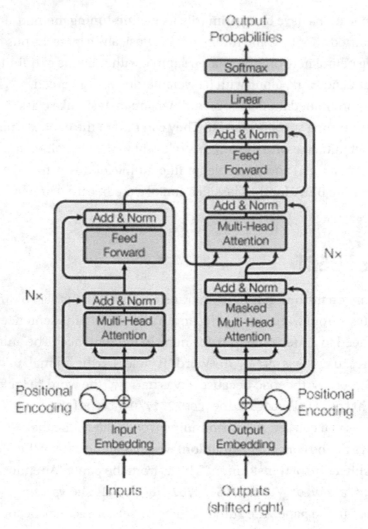

***Figure 5-21.*** *Transformer architecture*

For the purpose of BERT, I will get into detail only on the encoder part since BERT uses only the encoder part of the transformer. The most important component of the encoder part of the transformer is the multi-head attention. To understand multi-head attention, you must first start with understanding self-attention. Self-attention is a mechanism that relates how different words in the sentences influence each other. Let's understand how attention works. The image in Figure 5-22 comes from the "Breaking BERT Down" article (https://towardsdatascience.com/breaking-bert-down-430461f60efb), which explains a simple way to understand the inner mechanism of self-attention. There are three vectors that are important to know. For each word, these three vectors are calculated during the training process:

1.  Query vector

2.  Key vector

3.  Value vector

In order to calculate the vectors, you initialize weight matrices: $WQ$, $WK$, $WV$. These weight matrices will be embedding dimension (let's say 512) cross dimension of the query, key, and value vectors (say 64). Weight matrices will be of dimension 512 * 64 in this case. See Figure 5-22.

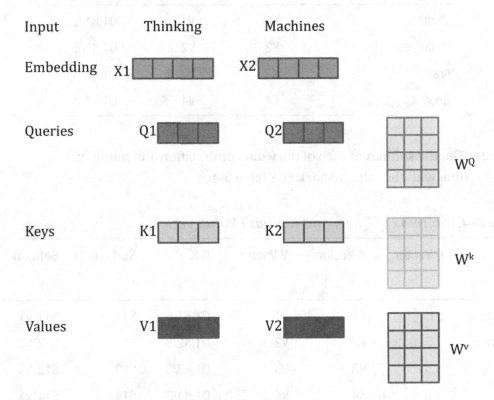

***Figure 5-22.*** *Query, key, and value vectors*

In the case shown in Figure 5-22, for the word "thinking" you get Q1 by multiplying the embedding of "thinking" with $W_Q$. Similarly, you get K1, V1 by multiplying with $W_{k,}$ $W_V$. The same is repeated for the word "Machines" using the embedding of "Machines."

Suppose you have the sentence "Thinking machines are good." The multiplications and further computation steps are represented in Table 5-3. The illustration is from www.analyticsvidhya.com/blog/2019/06/understanding-transformers-nlp-state-of-the-art-models/?utm_source=blog&utm_medium=demystifying-bert-groundbreaking-nlp-framework.

1. Multiply the query vector of the word with all other key vectors of other words. Divide that by the dimension of the square root of the query vector (8, square root of 64).

**Table 5-3.** *Query and Key Vector Dot Product*

| Word | Q Vector | K Vector | V Vector | Q.K |
| --- | --- | --- | --- | --- |
| Thinking | Q1 | K1 | V1 | Q1.K1/8 |
| Machines | | K2 | V2 | Q1.K2/8 |
| Are | | K3 | V3 | Q1.K3/8 |
| Good | | K4 | V4 | Q1.K4/8 |

2. Get the softmax of each of the word combination and multiply that with the value vectors. See Table 5-4.

**Table 5-4.** *Softmax of Query and Key with Value Vectors*

| Word | Q Vector | K Vector | V Vector | Q.K | Softmax | Softmax* Val |
| --- | --- | --- | --- | --- | --- | --- |
| Thinking | Q1 | K1 | V1 | Q1.K1/8 | S11 | S11*V1 |
| Machines | | K2 | V2 | Q1.K2/8 | S12 | S12*V2 |
| Are | | K3 | V3 | Q1.K3/8 | S13 | S13*V3 |
| Good | | K4 | V4 | Q1.K4/8 | S14 | S14*V4 |

3. Sum up all the vectors to get a vector that represents the given word into a Z matrix. Repeat the process for the other words. See Table 5-5.

**Table 5-5.** *Summing Up Vectors*

| Word | Q Vector | K Vector | V Vector | Q.K | Softmax | Softmax* Val | Sum up |
|------|----------|----------|----------|-----|---------|--------------|--------|
| thinking | Q1 | K1 | V1 | Q1.K1/8 | S11 | S11*V1 | Z1 = S11*V1 + ...S14*V4 |
| Machines |  | K2 | V2 | Q1.K2/8 | S12 | S12*V2 | |
| Are |  | K3 | V3 | Q1.K3/8 | S13 | S13*V3 | |
| Good |  | K4 | V4 | Q1.K4/8 | S14 | S14*V4 | |

The Z1 here is a vector of length 64. Similarly, you get vector of length 64 for all the words. I explained the case of single-head attention. In case of multi-head, you initialize multiple sets of Q, K, V matrices for each word embedding (by initializing multiple sets of W$Q$, W$K$, W$V$). A quick illustration in shown in Figure 5-23.

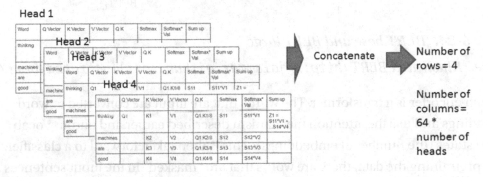

**Figure 5-23.** *Multi-head attention*

You get multiple Z outputs (let's say you have four attention heads) and then you have a 4*64 matrix at each head. All of them get concatenated with one another, the number of rows being 4 (the sentence and vocab have only 4 words) and the columns being the dimension of Q,K,V vectors, which is 64 multiplied by the number of heads (in your case its 4). This matrix further gets multiplied by a weight matrix to produce output of the layer. The output of the transformer is the number of words * 64 in this case.

The BERT architecture on the outset can be represented in Figure 5-24 in two variations.

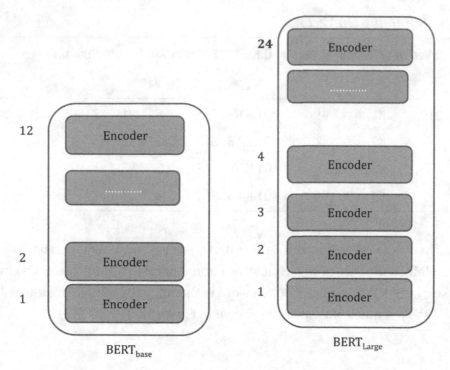

***Figure 5-24.*** *BERT base and BERT large*

*Source:  Illustrated BERT (http://jalammar.github.io/illustrated-bert/)*

Each encoder is a transformer. The words that are input are converted to word
embeddings and use the attention mechanism described earlier and output Vocab *
hidden states (the number of embeddings). This is then taken forward to a classifier.
While pretraining the data, there are words that are "masked" in the input sentences
and the classifier trains to identify the masked word. The BERT base has 12 layers
(transformer/encoder blocks), 12 attention heads, and 768 hidden units while BERT large
has 24 layers (transformer/encoder blocks), 16 attention heads, and 1024 hidden units.

# Fine-Tuning BERT for a Classifier

You will retrain the BERT base for your classifier use case. For this, the data has to be
converted into a BERT-specific format. You need to tokenize the words based on BERT
vocabulary. You also need to have positional tokens as well as sentence segment tokens.
Positional tokens are positions of the word and segment tokens represent the position of the
sentences. These layers form the input to the BERT and then get fine-tuned in the last few
layers using the loss-function for multi-class. An example of the input is given in Figure 5-25.

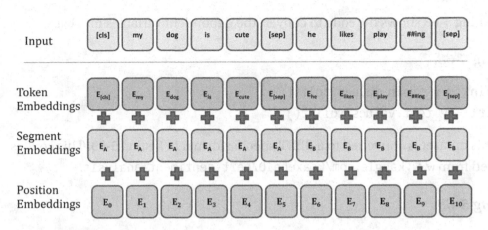

*Figure 5-25.   BERT inputs*

*Source:  Analytics Vidhya (www.analyticsvidhya.com/blog/2019/09/ demystifying-bert-groundbreaking-nlp-framework/)*

You now use a library called ktrain (see "ktrain: A Low-Code Library for Augmented Machine Learning" at https://arxiv.org/abs/2004.10703) to fine tune your BERT model for the credit card dataset problem. You divide the credit card dataset (dataframe t1) into train and test datasets herewith. You will be using Google Collab for the same. Listing 5-66 installs the packages that you need to run BERT using ktrain. See also Listing 5-67.

*Listing 5-66.*

```
!pip install tensorflow-gpu==2.0.0
!pip install ktrain
```

*Listing 5-67.*

```
tgt = t1["Final intent"]

from sklearn.model_selection import StratifiedShuffleSplit
sss = StratifiedShuffleSplit(test_size=0.1,random_state=42,n_splits=1)

for train_index, test_index in sss.split(t1, tgt):
x_train, x_test = t1[t1.index.isin(train_index)], t1[t1.index.isin
(test_index)]
y_train, y_test = t1.loc[t1.index.isin(train_index),"Final intent"],
t1.loc[t1.index.isin(test_index),"Final intent"]
```

Listing 5-68 shows the code to convert the dependent variables into lists.

**Listing 5-68.**

```
y_train1 = y_train.values.tolist()
y_test1 = y_test.values.tolist()
```

This dataset is converted to a BERT-specific format in Listing 5-69. This code is referred from www.kaggle.com/ksaxena10/bert-sentiment-analysis.

**Listing 5-69.**

```
from ktrain import text
import numpy as np
(x_train2,  y_train2), (x_test2, y_test2), preproc = text.texts_from_
array(x_train=np.array(x_train["Line"]), y_train=y_train1,

        x_test=np.array(x_test["Line"]), y_test=y_test1,

        class_names=class1,

        preprocess_mode='bert',

        ngram_range=1,

        maxlen=350)
```

The data is tokenized using BERT tokenization and numbers assigned for the given sequence length. Using preproc = "BERT" we tell the format that is needed to be kept. The sentence is also broken into segments and provided segment ids. See Listing 5-70 and Figure 5-26.

**Listing 5-70.**

```
x_train[0].shape,x_train[1].shape
((500, 350), (500, 350))

x_train[0][1]
```

```
array([ 101, 1045, 2572, 5307, 3314, 2007, 2026, 4923, 4003,  102,    0,
          0,    0,    0,    0,    0,    0,    0,    0,    0,    0,    0,
          0,    0,    0,    0,    0,    0,    0,    0,    0,    0,    0,
          0,    0,    0,    0,    0,    0,    0,    0,    0,    0,    0,
          0,    0,    0,    0,    0,    0,    0,    0,    0,    0,    0,
          0,    0,    0,    0,    0,    0,    0,    0,    0,    0,    0,
          0,    0,    0,    0,    0,    0,    0,    0,    0,    0,    0,
          0,    0,    0,    0,    0,    0,    0,    0,    0,    0,    0,
          0,    0,    0,    0,    0,    0,    0,    0,    0,    0,    0,
          0,    0,    0,    0,    0,    0,    0,    0,    0,    0,    0,
          0,    0,    0,    0,    0,    0,    0,    0,    0,    0,    0,
          0,    0,    0,    0,    0,    0,    0,    0,    0,    0,    0,
          0,    0,    0,    0,    0,    0,    0,    0,    0,    0,    0,
          0,    0,    0,    0,    0,    0,    0,    0,    0,    0,    0,
          0,    0,    0,    0,    0,    0,    0,    0,    0,    0,    0,
          0,    0,    0,    0,    0,    0,    0,    0,    0,    0,    0,
          0,    0,    0,    0,    0,    0,    0,    0,    0,    0,    0,
```

*Figure 5-26.*

Tokenized words are shown in Listing 5-70. From here it is 3 lines of code in the model. We pass 'bert' as a parameter to "text_classifier" so that the BERT model is fine tuned. The preproc parameter auto generates the format that "BERT" requires. We use a learning rate "2e-5" as it is the recommended learning rate. See Listing 5-71 and Figure 5-27.

*Listing 5-71.*

```
model = text.text_classifier('bert', train_data=(x_train, y_train),
preproc=preproc)
learner = ktrain.get_learner(model, train_data=(x_train, y_train),
batch_size=6)
hist = learner.fit_onecycle(2e-5, 2)
```

```
begin training using onecycle policy with max lr of 2e-05...
Train on 500 samples
Epoch 1/2
500/500 [==============================] - 2078s 4s/sample - loss: 0.4629 - accuracy: 0.8240
Epoch 2/2
500/500 [==============================] - 2054s 4s/sample - loss: 0.0303 - accuracy: 0.9960
```

*Figure 5-27.*

As you can see, the training accuracy is 99.6%. In Listing 5-72, you execute the model on a sample sentence.

*Listing 5-72.*

```
predictor = ktrain.get_predictor(learner.model, preproc)
predictor.predict(['waive my annual fee'])
['Annual Fee Reversal']
```

# Further Nuances in Building Conversational Bots

## Single-Turn vs. Multi-Turn Conversations

Figure 5-28 shows examples of single-turn conversations. Single-turn conversational bots are successfully used in many organizations. As you can see, this is a question/ answer pair. If you get the intent of the customer right, you can get the response right.

| Customer | What is my balance |
|---|---|
| Bot | Your account balance as of 3rd May is $XXX. |
| Customer | Where can I find the nearest ATM. |
| BOT | Checking from your current location, your nearest ATMs are listed in this link. |

*Figure 5-28.* *Single-turn conversations*

However, in multi-turn conversations there are multiple question/answer pairs. The bot needs to understand the "intent" of the user along with some more entity level meta data. For example for a restaurant booking - the bot needs to understand that it is a reservation related question (intent) along with the place, cuisine, number of persons, timing prefernces of the user (entities). These information needs to be collected before the bot can proceed to resolve the issue. The user sometimes may state the problem plainly ("i want to book a table") or with partial information ("I want to a book a table at XYZ restaurant for 4 people"). The bot may need to understand a set of entity values and the user may have stated only a part of the set. The bot then identifies the missing information and collects those. See Figure 5-29.

| Bot | How can i help you? |
| --- | --- |
| Customer | Want to book a table for 2 |
| Bot | Sure. Which location are you looking at? |
| Customer | I stay in east of avenue street and hence looking for places close by |
| Bot | Which Cuisine would you prefer? |
| Customer | Fine Dine Indian Restaurant |
| Bot | Great to know that. I can book table for 2 at Punjabi Frill at 8PM this evening. |
| Customer | Perfect |
| Bot | Have a great day. It's a pleasure to serve you. |

***Figure 5-29.*** *Multi-turn conversations*

In the example in Figure 5-29, if the user says "Want to book a table for 2 at east of avenue street," then the bot should be "smart" enough to skip the location question and ask the cuisine question. If all information is already provided, it should directly proceed to booking. This is achieved by keeping slots for different entities in intents and updating the slots as the conversation goes. If the slot is already filled, the question is skipped.

Sometimes in conversation, the user may not answer the question and change the intent or go back and correct the answer of a previous question. An example is given in Figure 5-30.

| Bot | How can i help you? |
| --- | --- |
| Customer | Want to book a table for 2 |
| Bot | Sure. Which location are you looking at? |
| Customer | I stay in east of avenue street and hence looking for places close by |
| Bot | Which Cuisine would you prefer? |
| Customer | **I think i want to change the location** |
| Bot | I see there is a change of location. |
| Customer | Yes |
| Bot | Which location would you prefer? |
| Customer | **I have some change of plans. Will call later** |
| Bot | Thanks. Have a great day. |

***Figure 5-30.*** *Intent switch*

In the "bold" lines the user has changed their intent or provided invalid responses. The bot workflow should be smart enough to change these responses. This can also be achieved using the encoder-decoder architecture where you are able to train a large corpus of pair-wise responses.

# Multi-Lingual Bots

An aspirational area for conversational agents to get into is multi-lingual bots. Given that the English-speaking population is only 20% of the world population, bots need to adapt to many languages. This poses a challenge since the research work in multilingual NLP is still very limited. However solving for vernacular bots are key to the success of natural language bots globally.

Human - Machine interaction has been a long felt need. We have seen some remarkable achievements in the recent past. We are no more fascinated by a machine that talks or replies the question posed. Soon enough we would not even recognize its a machine that we are talking to and treat it like any other interaction in our daily life.

Table 5-6 lists the packages used in this chapter.

***Table 5-6.***  *List of Packages*

| pack_name | version |
| --- | --- |
| absl-py | 0.7.1 |
| astor | 0.8.0 |
| attrs | 19.1.0 |
| backcall | 0.1.0 |
| bleach | 3.1.0 |
| blis | 0.2.4 |
| boto3 | 1.9.199 |
| boto | 2.49.0 |
| botocore | 1.12.199 |
| certifi | 2019.3.9 |
| chardet | 3.0.4 |

*(continued)*

***Table 5-6.*** (*continued*)

| pack_name | version |
| --- | --- |
| click | 7.1.2 |
| colorama | 0.4.1 |
| cycler | 0.10.0 |
| cymem | 2.0.2 |
| dataclasses | 0.7 |
| decorator | 4.4.0 |
| defusedxml | 0.6.0 |
| docopt | 0.6.2 |
| docutils | 0.14 |
| eli5 | 0.9.0 |
| en-core-web-md | 2.1.0 |
| en-core-web-sm | 2.1.0 |
| entrypoints | 0.3 |
| fake-useragent | 0.1.11 |
| filelock | 3.0.12 |
| fuzzywuzzy | 0.18.0 |
| gast | 0.2.2 |
| gensim | 3.8.0 |
| graphviz | 0.11.1 |
| grpcio | 1.21.1 |
| h5py | 2.9.0 |
| idna | 2.8 |
| imageio | 2.9.0 |
| inflect | 2.1.0 |
| ipykernel | 5.1.1 |

(*continued*)

***Table 5-6.*** (*continued*)

| pack_name | version |
|---|---|
| ipython-genutils | 0.2.0 |
| ipython | 7.5.0 |
| jedi | 0.13.3 |
| jinja2 | 2.10.1 |
| jmespath | 0.9.4 |
| joblib | 0.13.2 |
| jsonschema | 3.0.1 |
| jupyter-client | 5.2.4 |
| jupyter-core | 4.4.0 |
| keras-applications | 1.0.8 |
| keras-preprocessing | 1.1.0 |
| keras | 2.2.4 |
| kiwisolver | 1.1.0 |
| ktrain | **0.19.3** |
| lime | 0.2.0.1 |
| markdown | 3.1.1 |
| markupsafe | 1.1.1 |
| matplotlib | 3.1.1 |
| mistune | 0.8.4 |
| mlxtend | 0.17.0 |
| mock | 3.0.5 |
| murmurhash | 1.0.2 |
| nbconvert | 5.6.0 |
| nbformat | 4.4.0 |
| network | 2.4 |

(*continued*)

***Table 5-6.*** (*continued*)

| pack_name | version |
|---|---|
| nltk | 3.4.3 |
| num2words | 0.5.10 |
| numpy | 1.16.4 |
| oauthlib | 3.1.0 |
| packaging | 20.4 |
| pandas | 0.24.2 |
| pandocfilters | 1.4.2 |
| parso | 0.4.0 |
| pickleshare | 0.7.5 |
| pillow | 6.2.0 |
| pip | 19.1.1 |
| plac | 0.9.6 |
| preshed | 2.0.1 |
| prompt-toolkit | 2.0.9 |
| protobuf | 3.8.0 |
| pybind11 | 2.4.3 |
| pydot | 1.4.1 |
| pyenchant | 3.0.1 |
| pygments | 2.4.2 |
| pyparsing | 2.4.0 |
| pyreadline | 2.1 |
| pyrsistent | 0.15.2 |
| pysocks | 1.7.0 |
| python-dateutil | 2.8.0 |
| pytrends | 1.1.3 |

(*continued*)

***Table 5-6.*** (*continued*)

| pack_name | version |
|---|---|
| pytz | 2019.1 |
| pywavelets | 1.1.1 |
| pyyaml | 5.1.1 |
| pyzmq | 18.0.0 |
| regex | 2020.6.8 |
| requests-oauthlib | 1.2.0 |
| requests | 2.22.0 |
| s3transfer | 0.2.1 |
| sacremoses | 0.0.43 |
| scikit-image | 0.17.2 |
| scikit-learn | 0.21.2 |
| scipy | 1.3.0 |
| sentencepiece | 0.1.92 |
| setuptools | 41.0.1 |
| shap | 0.35.0 |
| six | 1.12.0 |
| sklearn | 0 |
| smart-open | 1.8.4 |
| spacy | 2.1.4 |
| srsly | 0.0.6 |
| stop-words | 2018.7.23 |
| tabulate | 0.8.3 |
| tensorboard | 1.13.1 |
| tensor-gpu | **2.0.0** |
| tensorflow-estimator | 1.13.0 |

(*continued*)

***Table 5-6.*** (*continued*)

| pack_name | version |
|-----------|---------|
| tensorflow | 1.13.1 |
| termcolor | 1.1.0 |
| testpath | 0.4.2 |
| textblob | 0.15.3 |
| thinc | 7.0.4 |
| tifffile | 2020.7.22 |
| tokenizers | 0.7.0 |
| tornado | 6.0.2 |
| tqdm | 4.32.1 |
| traitlets | 4.3.2 |
| transformers | 2.11.0 |
| tweepy | 3.8.0 |
| typing | 3.7.4 |
| urllib3 | 1.25.3 |
| vadersentiment | 3.2.1 |
| wasabi | 0.2.2 |
| wcwidth | 0.1.7 |
| webencodings | 0.5.1 |
| werkzeug | 0.15.4 |
| wheel | 0.33.4 |
| wincertstore | 0.2 |
| word2number | 1.1 |
| wordcloud | 1.5.0 |
| xgboost | 0.9 |

# Index

## A

Adpositional phrase (ADP), 126
Agent attributes, 52
Anchor words, 89–91, 93
Applying model
    negative corpus, 166
    NER dataset, 159
    only_ner_tagged.csv, 165
    positional variables, 161
    predicted organization, 163
    string and number columns, 162
    tokenized inputs, 163
    word level, 164
    word type variables, 160
    x_test dataset, 162
ax.twinx() function, 48

## B

Banking, financial services, and insurance
    industries (BFSI), 121
    concatenating columns, 122
    natural language generation, 177
    NLP data, 177
    risks, 121
    sample dataset, 122
    SMSdata, 177
Bidirectional Encoder Representations
    from Transformers (BERT), 187
    architecture, 232
    classifier use cases, 236–239

language models/fine-tuning, 230, 231
    multi-head attention, 235
    query, key and value vectors, 233, 234
    summing up vectors, 235
    training process, 233
    transformers, 231
Bidirectional LSTM
    decoder, 228–230
    encoder, 228
    training model, 227
Booster words, 73
Building Conversational Bots
    intent switch, 242
    multi-lingual bots, 242–247
    single-turn *vs.* multiple-turn
        conversations, 240, 241

## C

Capitalization, 86
CBOW
    binary classifier, 136
    200-dimensional vector, 139
    gensim library, 136, 137
    iter option, 138
    negative sampling, 136
    network architecture, 136
    skip gram model, 135
    target word, 134
    word embeddings, 139
    Word2Vec, 137

249

Printed in the United States
By Bookmasters